TrapGPT

the Hustlas Guide to AI

For my mother. Fifth-grade education, doctorate in courage. She passed when I was nineteen, but her upbringing kept raising me, year after year, toward discipline, study, and service. Everything I learn I share; everything I share traces back to her. I offer this book in gratitude and proof.

Table of Contents

Welcome to the New Hustle
AI as Your Silent Partner

Welcome to the future. But hold up; this isn't the future they promised, filled with flying cars and robots taking our jobs. Nah, this is the future we've built from the ground up, where street smarts meet digital hustle. You're here because you know the game is changing, and you refuse to get left behind.

Think of AI as the ultimate silent partner; always awake, always learning, and always ready to put in work. This partner doesn't sleep, doesn't complain, and definitely doesn't make excuses. It's like having the most reliable plug you've ever known, one that only speaks success, efficiency, and straight-up results.

Let's keep it real. As minority entrepreneurs, we've always had to grind twice as hard, face twice the barriers, and navigate twice the risks. But guess what? AI doesn't care about where you come from or what you look like. It cares about how effectively you can prompt, command, and leverage it to stack your success.

In this book, we're flipping the script. We're using AI not to replace our hustle, but to elevate it. Whether your brand new to AI or already dabbling, this guide is your blueprint to mastering the digital come-up.

Ready to level up your hustle? Let's go.

Why Minority Entrepreneurs Need to Master AI

Look, we're no strangers to disruption. As minority entrepreneurs, we've always known how to hustle smarter and grind harder. But here's the truth: the game has changed. The hustle has evolved, and now AI is rewriting the rules. Either we master this new wave or risk being sidelined.

Historically, minority businesses have faced relentless barriers, limited access to funding, fewer networking opportunities, systemic biases, and the ever-present need to stretch every dollar. These challenges haven't disappeared, but AI presents a powerful tool to overcome them, leveling a playing field that was never built for us.

By mastering AI, you amplify your hustle exponentially. Tasks that once drained your resources like legal paperwork, marketing strategies, bookkeeping, even creative content, can now be done quickly, efficiently, and at a fraction of the traditional cost. AI is no longer just a buzzword reserved for big tech; it's a lifeline for the streets, a democratizer of innovation, and the ultimate empowerment tool.

The reality is simple: if you don't adapt, someone else will. Mastering AI isn't optional, it's essential. Embracing AI technology means stepping boldly into spaces once denied to us, not asking for permission, but claiming what's rightfully ours.

We don't follow trends; we set them. It's time to take AI, flip it, and make it serve your vision, your dreams, and your community.

AI mastery isn't just your competitive edge, it's your birthright. Let's claim it.

How This Book Will Change Your Business (and Life)

This isn't your typical business book. It's a movement, a toolkit, and a blueprint designed specifically for entrepreneurs who know the grind all too well. By the time you finish reading, your entire approach to business will have shifted. The stress of doing everything yourself, of stretching your resources paper-thin, will be replaced by clarity, efficiency, and a renewed sense of purpose.

Each chapter of *TrapGPT: The Hustla's Guide to AI* is crafted to show you exactly how AI tools can elevate every part of your hustle. You'll learn how to communicate with AI clearly and powerfully through prompts, saving countless hours and dollars along the way. From handling tedious office tasks to navigating complex legal issues, creating engaging ads, and building a brand that resonates with your community, AI can handle it all. You'll even learn to automate your finances, track your money smarter, and discover opportunities you never knew existed.

This isn't about replacing your hustle; it's about enhancing it. Imagine having more freedom to innovate, strategize, and lead because you're no longer bogged down by tasks that technology can handle. Think about the confidence you'll gain knowing you're leveraging the same tools that big corporations use, but doing it your way, for your vision.

You won't just build a stronger business; you'll reclaim your time, your creativity, and your peace of mind. This book is about transformation. It's about going from just surviving in the game to thriving, using AI to ensure that your hustle outlasts trends, competitors, and even your own expectations.

Ready to take back control? This journey is yours and it begins right here.

Chapter One

99 Problems
but a Prompt Ain't 1

If AI is the engine driving the new hustle, prompts are the keys you turn to get it started. Without the right prompts, even the most powerful AI is just a fancy piece of software collecting digital dust. But master prompting, and you've unlocked the power to automate, innovate, and elevate your business beyond your competition's reach.

Think of prompts as conversations. Clear instructions you give an AI tool to generate exactly what you need. Whether you're drafting a contract, building a marketing plan, or creating viral content, the quality of your prompts determines the quality of your results. Weak prompts lead to weak outcomes, wasting valuable time and money. Strong prompts deliver precision, efficiency, and results tailored specifically to your hustle.

Here's the trap truth: Most people underestimate the power of prompts. They see it as simple text entry. Typing out a few words and hoping for the best. But elite hustlers know better. They understand that prompting is an art form and a strategic tool that separates casual users from true AI masters.

By learning how to craft effective prompts, you place yourself ahead of the game. You start using AI not just to keep up, but to dominate. This book will teach you exactly how to do that beginning with mastering prompts.

Ready to learn the secrets of prompting like a pro? Let's dive deeper and start speaking the AI language your hustle demands.

The Prompt Framework: *Persona, Task, Context, Critique, Refine*

Prompting isn't random, it's strategic. When done right, it transforms vague requests into powerful results. To unlock maximum value from any AI tool, you need a structured approach. Enter the Prompt Framework: a proven method that breaks down prompt creation into five critical components. Persona, Task, Context, Critique, and Refine.

1. Persona (Tell the AI Who or What to Be)

Before the AI does anything for you, you gotta set the scene. You're not just typing to a robot; you're talking to a role-player. That's what *Persona* is all about: telling the AI who it needs to become to get the job done right.

Think of it like hiring the right person for the job. If you need legal advice, you're not asking your barber. If you're designing a logo, you're not calling your accountant. You need to *cast the role* before the performance begins. The more specific you are, the better the AI steps into character.

With Persona, you give the AI a hat to wear, a role to play. The vibe, the expertise, the energy. If you need a hype-beast brand strategist? Say that. A grant writer with experience helping Black-owned non-profits? Say that too. Give it credentials. Give it purpose.

Basic Prompt:

> *You are a seasoned business attorney who specializes in minority-owned startups.*

Enhanced Prompt with flavor:

> *You're a no-nonsense, street-smart attorney with 15 years of experience protecting Black and Brown entrepreneurs from getting played. You break down legal language in a way that anyone from the barbershop to the boardroom can understand.*

See the difference? The second one brings character. Tone. Depth. It doesn't just tell the AI *what* to do; it tells it *how* to show up.

Why Persona Matters:

This is what sets apart amateur prompting from elite-level prompting. The right persona ensures your results aren't just correct; they're aligned with *your culture, your audience,* and *your mission.* You don't want just answers. You want answers that sound like they came from your circle, not a corporate cubicle.

So, when in doubt, treat the AI like your new employee: tell it who it is, give it a role, and set it up to win for you.

This ain't just prompt work. This is casting calls for digital greatness.

2. Task (Tell the AI What the Play Is)

Once you've set the persona, the next move is simple, but critical: *tell the AI exactly what you want it to do.* Don't beat around the bush. Don't leave it up to interpretation. In the AI game, clarity is currency.

The *Task* is the actual job. It's the hustle you're delegating. Whether it's writing a business proposal, drafting a cease-and-desist letter, or laying out a month-long content plan; it starts with your direction.

Here's where most people mess up: they ask AI like they're texting a friend. "Can you help me write something?" Nah, fam. That's not how you move with purpose. Be specific. Be surgical. This ain't a guessing game, this is business.

Weak Task Prompt:

> *Write something about marketing.*

Strong Task Prompt:

> *Create a 4-week marketing plan that includes weekly themes, email subject lines, and social media post ideas for a Black-owned skincare brand targeting millennials.*

See the difference? One is a shot in the dark. The other is the blueprint.

Thanks Hov!

The Hustla's Rule:

"If you don't know exactly what you want, the AI won't either."

You gotta drive the car, you can't expect it to guess the destination. The tighter your task, the better your outcome.

8

Here's a cheat code: start your task prompts with action verbs.

Words like: **Create. Draft. Write. Develop. Design. Explain. Break down. Map out.**

These tell the AI, "This is the job. Do it now. Do it right."

The Hustla's Reminder:

"AI is your assistant, not your mind reader. You wouldn't hand a new intern vague instructions and expect excellence. So don't do it here either. You've got the persona in place, now give them their mission."

In the world of AI prompting, the task *is the play.* Call it clearly. Execute precisely. And watch how the AI runs it like clockwork.

3. Context (Set the Scene, Feed the Machine)

Now that you've told the AI *who it is* and *what to do*, the next key is giving it *why* and *how*. That's where **context** comes in.

Context is the background, the vibe, the details that make your task make sense. It's the information that helps the AI understand your audience, your goals, your tone, your industry and everything that makes your request unique to *your* business.

AI is smart, but it's not psychic. If you leave it in the dark, it'll give you cookie-cutter answers. But when you shine a light, feeding it the right details, you get output that's laser-focused, on-brand, and ready to move the crowd.

Here's how to build strong context:

- Who is your audience?
- What's the goal or objective?
- What platform or format is it for?
- What tone or style should it match?
- Any special info it needs to know?

Example:

"Target audience: Black-owned small businesses in the beauty industry. Goal: Increase newsletter signups through Instagram content. Tone: Empowering, professional with a touch of humor. Platform: Instagram Reels and Stories."

Layer that context into your prompt and suddenly your AI isn't just guessing, it's performing.

Weak Context:

"Make it catchy and fun."

Strong Context:

"Write in a voice that feels like Nipsey Hussle meets Shark Tank. Sound authentic, powerful, and built for the culture. This is for Black Gen Z creatives who don't like corporate speak."

The Hustla's Reminder:

"The streets taught us: you gotta know the environment to survive in it. Same with AI. Context is your way of setting the environment, so the response thrives in your world."

So feed the machine. Put it on game. And let it give you results that don't just sound good, they sound like *you*.

4. Critique (Tell It What Not to Do and Check the Work Yourself)

By now, you've built a solid prompt. You've set the persona, clarified the task, laid down the context. Now it's time to draw the line; **what not to do** and, more importantly, to **check the work like a boss**.

This is *Critique*.

Let's keep it a buck: AI is powerful, but it ain't perfect. It moves fast, but sometimes it moves wrong. That's why critique isn't just about telling the AI what to avoid, it's also about **putting your eyes on the final product and holding it accountable**. You wouldn't let a new employee send out work without reviewing it. Don't treat AI any different.

Even the best prompt can return a half-right response. Maybe the facts are slightly off. Maybe the tone misses the mark. Maybe it's just giving *basic* when you needed *brilliant*.

That's your cue to evaluate and revise.

Here's how you check the work:

- Read it *out loud*. Does it sound like you?
- Cross-check any claims or data. AI will confidently make things up.
- Check for tone and audience alignment. Does this speak to your people?
- Ask: *Would I post this with my name on it?* If the answer's no, tweak it.

Critique = Quality Control.

Don't just copy and paste. Copy, **pause**, and paste with pride. Make edits. Tighten the message. Add your flavor. AI gets you 80% of the way, but *you* bring the sauce that seals the deal.

The Hustla's Reminder:

"AI is the tool. You're the talent."

Use the machine but trust your gut. Check the work. Protect your brand. Never hand over your voice without reading what it's saying in your name.

5. Refine (Run It Back 'til It's Right)

This is where we separate the hustlas from the hobbyists. Real talk, *Refine* is your chance to elevate. You've given the AI your prompt, but that first draft? It's just a starting point. Good output is cool. But **great** output? That comes when you *run it back*.

Refine means you don't just take what the AI gives you and keep it pushing. Nah. You **ask better follow-ups**, challenge the response, and push it to do more. Add more emotion. Make it clearer. Shorten it. Expand it. Translate it to your tone.

This ain't a one-and-done game. It's *iteration with intention* until what you see on the screen matches the vision in your head.

Here's how you refine:

- "Say it with more passion."

- "Now rewrite that in under 100 words for Instagram."
- "Give me three options with different tones: bold, inspirational, and funny."
- "Turn this outline into a script."
- "Now make it sound like it came from a hustla in Atlanta with something to prove."

Refine = Reps.

Think of AI like a gym. The more reps you do, the more feedback you give, the stronger the final result. Don't be lazy with it. Challenge it. Guide it. Mold it. Because the second, third, or fourth version? That's usually where the gold lives.

And yo, don't forget *refining isn't just for content.* It works for proposals, policies, contracts, captions, everything. Every draft is a brick. But you're building a whole house.

The Hustla's Rule:

"The first draft is the alley-oop. The final draft is the dunk. Don't settle. Refine until it hits different. Until it sounds like you. Until it moves like you. Until it speaks your brand's truth out loud."

Oh yeah, the AI's smart. But it ain't magic.

You don't just ask once and walk away like it's a genie in a bottle. Nah, this is more like working the stove; season, taste, adjust, repeat. Keep stirring the pot 'til the flavor hits. Because when it hits, **it hits**.

Refining is your moment to flex. It's where you show that you're not just using AI, you're *directing* it. You're the executive producer of every prompt, the creative director of every response, the editor-in-chief of your entire digital hustle.

You ain't here for basic. You're building legacy.

So, treat every AI output like a demo track. Play it back. Remix it. Master it. And when it finally slaps? Drop it like it's hot, because *that's* how real hustlas launch content, contracts, and campaigns.

The Hustla's Rule:

"The AI brings the beat. You bring the bars. Never be afraid to hit that "run it back" button 'cause greatness never drops on the first take."

Crafting Powerful Prompts: Step-by-Step Guide

You've got the framework. Now let's put it into motion and walk through how to actually build a prompt that gets you fire results, *not just decent ones.*

Here's your step-by-step guide to crafting prompts that don't miss:

Step 1: Define the Persona

Start by telling the AI *who it is*. Make it clear what role it's playing. Add personality, tone, even background experience if it helps.

Example:

"You are a culturally fluent brand strategist who specializes in helping Black-owned businesses find their voice."

Step 2: Clarify the Task

Be specific about what you want the AI to do. Give it a job with direction. Avoid vague phrases like "help me write something."

Example:

"Write a compelling brand mission statement."

Step 3: Provide the Context

Tell the AI *why* this matters and *who* it's for. Include tone, audience, industry, and platform info.

Example:

"This mission statement is for a vegan skincare line targeting Black women aged 25–35. The tone should be uplifting, luxurious, and rooted in self-love."

Step 4: Set the Boundaries (Critique)

Call out what *not* to do. This avoids generic, off-brand responses.

Example:

"Avoid overused phrases like "all-natural" or "beauty redefined." Don't sound like a generic ad."

Step 5: Refine It

Once the AI gives you a draft, push it to do better. Ask for tweaks, rewrites, or shorter versions. Don't settle on the first take.

Example:

"Now revise that to make it more emotionally powerful. Give me three variations with different tones: bold, poetic, and conversational."

The Hustla's Rule:

"A sloppy prompt gets you basic. A sharp prompt builds a business. Follow these steps, throw in those delimiters, and watch the AI work like it's got equity in your company."

Common Mistakes and How to Avoid Them

Even with the framework, even with the steps, folks still fumble the bag. Why? Because they rush the process. They treat prompting like casual Googling instead of precision work. Let's fix that. Here are the most common mistakes people make when using AI and how to dodge every single one.

✖ Mistake 1: Being Too Vague

What it looks like:

"Help me write something for my business."

Why it fails:

The AI has no idea what kind of business, who your audience is, or what you actually want. That's like walking into the barbershop and saying, "Just do something." Nah. Be specific. Use the full framework: Persona, Task, Context, Critique, Refine.

✖ Mistake 2: Skipping the Persona

What it looks like:

"Write a brand strategy."

Why it fails:

Without telling the AI *who* it's supposed to be, it gives you generic, vanilla responses. You want flavor. You want precision. So cast the right role first.

✖ Mistake 3: Ignoring Tone and Audience

What it looks like:

"Make a flyer."

Why it fails:

A flyer for a youth hip-hop event in Atlanta doesn't sound or look like one for a nonprofit gala in D.C. Always include context about *who* you're speaking to and *how* it should sound.

✖ Mistake 4: Copying and Pasting Without Reviewing

What it looks like:

Dropping the AI output straight into an email, ad, or contract without reading it.

Why it fails:

AI makes stuff up. It gets tone wrong. It misses details. You are the last line of defense. Always check the work. Edit it. Refine it. Put your name on it only when it's right.

✖ Mistake 5: Giving Up After the First Try

What it looks like:

"This AI thing doesn't work."

Why it fails:

It's not that the tool is broken, it's that you stopped too early. The first draft is *never* the best draft. Prompt. Review. Refine. Run it back until it hits.

The Hustla's Rule:

> *The AI is only as strong as the hand that's guiding it. If you treat it like a shortcut, you'll get shortcut results. But if you treat it like a weapon in your arsenal? You'll be unstoppable. Learn the game. Respect the process. And never settle for mid.*

Closing: Own the Code, Own the Game

This ain't just about writing better prompts, it's about rewriting your playbook.

The same way we learned to flip product, pitch vision, and move smart in rooms built to overlook us; this is that evolution. This is *digital dope dealing*, except now, the product is information. The grind? Automated. The results? Scalable. And the power? Back in your hands.

Mastering prompts means you ain't waiting on a team you can't afford. You just hired a squad of digital soldiers, 24/7, no days off, no ego, no payroll. You're the visionary *and* the operator. You're cutting costs, building systems, and freeing your time so you can scale, strategize, and shift the culture.

This chapter was the blueprint. The foundation. The first brick in your digital empire. Now you know how to talk to AI the right way, like a boss with direction, not a beginner throwing darts in the dark.

The old hustle was about muscle. This new hustle? It's about mastery.

The Hustla's Reminder:

"They taught us to code the streets. Now we code the machine."

And when you own the code?

You own the game.

TrapGPT Hustla Homework: Prompt Like a Pro

Now that you've got the blueprint, it's time to put in that work. You've learned the framework: **Persona, Task, Context, Critique, Refine**, and you've seen how powerful clean prompts can be. Now it's your turn to flip it.

🎯 **Your Mission:**
Write a prompt that tells an AI tool to help you with something *real* in your business.

This could be:
- Writing a product description
- Drafting an email to potential clients
- Creating a social media caption for your next post
- Generating ideas for a community event
- Outlining a business proposal

📌 **Your Prompt Must Include:**
☑ A clear **Persona**
☑ A well-defined **Task**
☑ Specific **Context** (audience, tone, goal)
☑ **Critique** to guide what to avoid
☑ A plan for how you'll **Refine** the AI's first draft

Example:

> *You are a street-smart marketing coach for Black-owned businesses. Write a caption for an Instagram Reel promoting a free AI for Entrepreneurs workshop. Audience: Minority small business owners aged 25–45 who want to scale. Tone: Bold, empowering, no fluff.*

Avoid: Overused phrases like "sky's the limit" or "level up." Keep it fresh.
Then, once the AI gives you the first draft:

👉 Ask it to improve it.
👉 Change the tone.
👉 Shorten or expand it.
👉 Or hit it with: *"Make it hit harder."*

Reflect:

After running the prompt, answer this:
- What did the AI get right?
- What did it miss?
- How did refining the output make it better?
- Would you use the final version in real life?

The Hustla's Rule:

"Practice builds precision. Precision builds profit."

You've got the playbook. Now run the play.

Chapter Two

Started From the Bottom
Let's Be Clear

Let's talk about the part they don't glamorize on Instagram. Running a business when you come from the mud ain't the same as running one with a trust fund behind you. There's no startup capital from Uncle Bob. No "friends and family" round. No cushion for mistakes. Just grind, pressure, and prayers.

When you're a minority entrepreneur, especially Black or Brown, the odds are stacked before you even open the doors. You're the CEO, the marketing team, the admin, the accountant, and the janitor all at once. And every move cost money you might not have yet.

You wanna launch? Cool.

- That's a few hundred for a logo.
- A few racks for a website.
- Thousands more for a lawyer to draft your contracts.
- Ads? Marketing? Copywriting? Automation? Most times, if you can't afford to outsource it, you just go without. Or you try to figure it out on your own... and sometimes, that's the beginning of the end.

Businesses don't always fail because the product is weak.

They fail because the *founder* is overworked, underfunded, and operating with no margin for error. No safety net. No room to scale, and that's where AI steps in, not as a luxury, but as a lifeline.

We're talking about using tech to cancel out some of the overhead that's been holding our people back for decades. We're talking about reclaiming hours, cutting costs, and still showing up like a professional even when it's just you behind the scenes.

Before we get into how AI solves the problem, we gotta understand how deep the problem runs. It ain't just about apps and algorithms. It's about access, survival, and freedom.

The Hustla's Reminder:

"The game is rigged. AI ain't the fix, but it's a weapon. Use it right, and you flip the script from underdog to unstoppable."

Death by a Thousand Invoices

Let's be real: most small businesses don't go broke off one big mistake. They bleed out slowly, invoice by invoice, month after month. A subscription here. A freelancer there. A lawyer for this. A consultant for that. It adds up fast. And if you're not watching close, your passion project becomes a pit that drains your time, your energy, and your wallet.

Here's what they don't tell you in the hype posts:

Running a business takes **admin**. It takes **systems**. It takes **documents**. That stuff ain't sexy, but it's necessary. And in the traditional world, all of it comes with a price tag.

Let's break it down:
- **Legal support?** $500 just for a basic contract. Thousands more if you catch a lawsuit.
- **Budgeting help?** That's a CPA on retainer or a software subscription with a learning curve.
- **Proposal writing?** $150–$500 per proposal if you outsource.
- **Marketing campaigns?** Creative agencies want four figures, minimum.
- **Social media?** You're paying someone monthly *or* spending 2–3 hours a day doing it yourself.
- **Automation tools?** Monthly fees. Consultant setup fees. Maintenance headaches.

All of these are real, recurring costs. And if you're bootstrapping or funding your business with your 9–5 paycheck, it's only a matter of time before you hit a wall. Either the money runs out or the energy does.

And here's the twist, **these are the very things that could grow your business if you could afford them.**

This is the trap. You don't grow because you can't invest. You can't invest because you're trying to stay afloat. And you're trying to stay afloat doing everything **by yourself.**

The Hustla's Reminder:

"It ain't always the product that fails. Sometimes it's the pressure that folds the dream."
But trust, there's a smarter way to carry the weight. And we're about to get into it.

Why AI Ain't Just a Cheat Code, It's the Equalizer

Listen. We ain't looking for handouts. Never have.

But what we do need is a way to close the gap and AI might just be the most powerful equalizer we've ever had access to.

This isn't some Silicon Valley toy.

This is a tool for the trenches.

A digital workhorse that don't sleep, don't call out sick, and doesn't need a 1099. It's here, it's accessible, and it's waiting on *you* to use it right.

You ever heard the saying "work smarter, not harder"?

This is what that actually looks like in practice.

Because here's the thing: AI can write that proposal.

It can generate those marketing captions.

It can help you draft that contract.

It can analyze your budget, plan your content, and automate your follow-ups.

Things you used to have to pay someone else for or burn yourself out trying to learn, are now sitting at your fingertips.

24/7. No ego. No hourly rate. Just results, and nah, it's not perfect. It's not gonna replace your hustle, but it's gonna *amplify* it. Multiply it. Make it stretch farther than it ever could on its own.

That's why this book exists.

Not to replace your grind, but to upgrade your gear. Because while the world's out here arguing about whether AI is "dangerous" or "the future," the real ones are already using it to *get free*.

The Hustla's Rule:

"Closed mouths don't get fed. And closed minds don't get funded."

AI is open and ready. You just gotta show up, type smart, and boss up.

The Freedom That Comes with Automation

We don't hustle just to stay busy; we hustle to buy freedom.

Time freedom. Mental freedom. Legacy freedom.

Too many of us stay stuck in survival mode because we're doing **everything ourselves** and convincing ourselves that makes us real.

Let's kill that myth right now.
Burnout ain't a badge of honor. Being the bottleneck in your own business ain't the move. Staying "busy" without growth? That's just a hamster wheel disguised as hustle.

Enter AI-powered automation; not just a tool, but a **time machine**, giving you back hours, peace, and clarity.

Imagine this:

- You wake up at 7 AM, coffee brewing, and your social media posts for the next two weeks are already written, scheduled, and queued. Each post tailored specifically for your audience, tone, and goals. While competitors spend mornings scrambling, you're strategizing next month's growth moves.

- A client emails asking for a detailed proposal ASAP. Instead of panicking, you input a quick prompt. Three minutes later, you have a polished, detailed proposal: formatted, professional, and laser-targeted for your client's specific needs and ready to send and secure the bag.

- New customer signs up online at 2 AM while you sleep. An automated sequence immediately sends a personalized welcome message, updates your CRM database, issues an invoice, and schedules their onboarding Zoom call. You wake up to confirmed appointments, completed paperwork, and cash in the bank without lifting a finger.

- Instead of losing Sunday nights buried in spreadsheets and receipts, AI-powered finance tools auto-update your books, highlight your most profitable products, pinpoint unnecessary expenses, and generate crystal-clear reports showing exactly where your dollars go and how to double them next quarter.

- You're invited to speak at a major conference. Normally, prepping slides and handouts would mean losing sleep. Instead, you drop a prompt in your AI tool, and 10 minutes later, you're handed a tight, dynamic, branded presentation complete with key points, visuals, and talking points that amplify your voice, your story, and your vision.

This isn't science fiction; this is how successful businesses move every single day. And now, thanks to AI, it's not reserved for million-dollar budgets. It's yours for the taking.

The more tasks you automate, the more time you have to innovate.

This ain't about doing **less**, it's about doing **more of what actually moves your business forward**. It's about elevating from hustler to CEO, from surviving to thriving, from grinding blindly to grinding strategically.

The Hustla's Reminder:

"The grind don't stop, but it can and better get smarter."

You didn't start your business to become an employee of your own hustle.

You built it to own something meaningful. Let AI handle the noise, so you can finally lead the movement.

Chapter Three

Legalize It
AI's Legal Game Strong

Let's talk about protection, not security guards, but paperwork., because **if your business ain't protected on paper, it ain't protected at all**. For too long, legal protection has been a gatekeeper. It's complex, expensive, and wrapped in a language that ain't made for us. Most of us didn't grow up around attorneys. We didn't get the blueprint on business formation, intellectual property, or contract negotiation. So we end up skipping legal steps, not because we're careless, but because we're under-resourced.

Here's the truth they don't say out loud:

- Most early-stage Black and Brown businesses are operating without proper contracts.
- Many of us never trademark our names or logos.
- We're relying on handshake deals, Word doc agreements, and legal templates pulled from random blogs.
- We hope nothing goes wrong, because we can't afford it if it does.

But hope ain't a strategy. And in today's economy, one legal misstep could cost you **everything** you've built.

Now here's the twist: **AI can actually help you protect your hustle,** and no, this isn't just hype, AI has *literally* passed the bar exam. It might not replace a full-time lawyer, but it can absolutely draft enforceable contracts, break down complex legalese into plain English, and help you move like a business that's ready for the big stage.

This chapter is about turning AI into your legal co-pilot. It's your entry-level legal team on call, on budget, and on your time.

We're going to show you how to use AI to:

- Draft airtight NDAs, contracts, and service agreements
- Write your own cease-and-desist letters with confidence
- Create terms of service and privacy policies for your website
- Understand legal clauses before you sign anything
- Get clear, step-by-step help with forming an LLC, partnership, or S-Corp

You won't just be guessing anymore. You'll be building with clarity, confidence, and control.

🧠 Sample Prompt: Drafting a Non-Disclosure Agreement (NDA)

Let's use the prompt framework from Chapter 1 to create a solid, AI-generated NDA.

Prompt Breakdown:

Persona:
You are a licensed business attorney who specializes in contracts for creative entrepreneurs and small business owners.

Task:
Draft a simple but legally sound Non-Disclosure Agreement (NDA).

Context:
This NDA is for a minority-owned media production company that's partnering with a freelance videographer on a new documentary series.

The NDA should ensure all shared information about the project remains confidential and protect proprietary content and footage.

Critique:
Avoid overly technical legal jargon that would be confusing to someone without a legal background. Make sure the agreement includes key elements like the definition of confidential information, duration of the agreement, and legal consequences if the NDA is broken.

Refine:
After the draft, reformat it for plain English readability, and provide a one-paragraph summary of what each section means in simple terms.

Full Prompt (to drop into your AI tool):

You are a licensed business attorney who specializes in contracts for creative entrepreneurs and small business owners. Draft a legally sound Non-Disclosure Agreement (NDA) for a minority-owned media production company that's working with a freelance videographer on a documentary project. The NDA should cover confidentiality of project details and footage, outline what constitutes confidential information, the duration of the NDA, and consequences for violation. Avoid dense legal jargon. Once you finish, rewrite the NDA in plain English with a summary of each section.

The Hustla's Rule:

"Every empire starts with paperwork."

You can't scale a brand that's legally exposed.

So, legalize it. Build your business on paper like you plan to pass it down. And let AI be the paralegal in your pocket while you move like a CEO.

Contract Season: *Lock It In, Lawyer Up, and Sue if you Need to*

Let's keep it all the way real:

You don't just need contracts to protect the bag. Sometimes, you need to **enforce them too.**

And other times? You need to *defend yourself* from folks trying to test your gangsta like you ain't built for this.

We already talked about how contracts are the foundation. They're the "nope, run me my money" document. The "we agreed to this in writing" receipt. But what happens *after* that when someone still crosses the line?

When you've got receipts, but now you need to take it to the next level? This is where most small business owners tap out. Because litigation feels too big. Too complex. Too expensive.

Hiring a lawyer to file a claim can cost thousands. Responding to a lawsuit you didn't see coming? Even more.

But here's what AI lets you do:

- Draft **basic legal complaints** to file in small claims court
- Create structured arguments and timelines to support your case
- Get help formatting and organizing your case documents
- Draft motions, responses, and summaries if you're **going pro se** (representing yourself)
- Understand legal procedures without needing a law degree

Let's Break It Down: Real-World Scenarios

Scenario 1: Filing a Lawsuit (Small Claims Court)

Let's say a vendor took your $2,500 deposit and ghosted. You've got the contract. You've got emails. They're not responding. You're done playing nice.

AI can help you:
- Draft a formal **legal complaint** that outlines what happened, what was agreed upon, and what laws were broken.
- Format the complaint based on your local court's requirements.
- Generate a **timeline of events** with supporting evidence.
- Write your **statement of claim** in a clear, legally sound format that you can file yourself.

You're walking into court with confidence not confusion.

Scenario 2: Defending Yourself Pro Se

Now flip the script.

Someone's coming at *you* with a claim. Maybe a disgruntled client. Maybe a former contractor. You can't afford a lawyer just yet, but you're not about to roll over either.

AI can help you:
- Break down their complaint into plain English so you understand what you're being accused of.
- Help you write a **response or answer to the complaint** with a structured legal defense.
- Draft a **counterclaim** if you've got one.
- Prepare statements, witness lists, and questions to use in your court appearance.
- Format motions or requests for extensions, discovery, or dismissal.

You're not just surviving the court process; you're **learning how to fight smart**.

You're showing up as a pro with limited resources, but *unlimited strategy*.

🖋 Sample Prompt: Filing a Small Claims Lawsuit

Persona:
You are a legal assistant with experience helping entrepreneurs file small claims lawsuits in local and state court.

Task:
Draft a complaint to file in small claims court against a marketing consultant who accepted payment but never delivered services.

Context:
The business owner paid $1,750 to a freelance consultant for a month-long social media ad campaign. The consultant missed deadlines, never launched the campaign, and stopped responding. The complaint should include the nature of the agreement, payment proof, timeline, attempts to resolve the issue, and the specific damages being requested.

Critique:
Avoid overly technical legal terms. Use language that's appropriate for court but accessible to a non-lawyer. Don't exaggerate claims, just lay out the facts clearly and assertively.

Refine:
After drafting the complaint, generate a checklist of what the business owner should bring to court (evidence, printouts, contract, emails, etc.) and a 60-second verbal summary of the case they can practice before appearing.

Pro Tip:

If you do end up hiring a lawyer later, this AI-drafted material makes their job easier and can lower your bill. It shows initiative, saves time, and ensures nothing gets lost in translation.

The Hustla's Rule:

"You don't need a million-dollar lawyer to protect a million-dollar vision."

Use AI to write the rules, defend the work, and if necessary, file the paperwork.

Because in this game, *paper protects power*. And power respects preparation.

Put It in Ink: *Contracts That Don't Fold Under Pressure*

In the streets or in the boardroom, real hustlas know: *if it ain't in writing, it ain't real.*

Verbal agreements sound good... until someone switches up. Memory gets fuzzy, fingers start pointing, and suddenly what was "understood" becomes a problem. You're not just building a brand; you're building an asset. That means you need receipts, structure, and documentation that can stand up when things get shaky.

That's where contracts come in.

A contract ain't just paperwork, it's your protection plan. Your boundary setter. Your business backbone. Whether you're hiring a videographer, bringing on a partner, licensing a design, or landing a paid speaking gig, the contract makes it clear: *here's what we agreed to, here's what happens if someone slips.*

Problem is, most small business owners don't have access to legal teams. You're bootstrapping, making it work, and legal is usually the last thing on the budget, until it's too late, but with AI in your corner, you don't have to choose between staying protected and staying afloat. You've now got a digital legal assistant that works 24/7, speaks every language, and doesn't charge by the hour.

What AI Can Help You Do with Contracts

☑ **Draft from Scratch**

Need a service agreement, partnership contract, or licensing deal? You can use AI to generate a full draft based on your industry, terms, and tone.

☑ **Customize Templates**

Already got a contract? Let AI review and revise it. Update the tone, change the terms, clarify the clauses fast.

☑ **Break Down Legal Jargon**

Don't sign what you don't understand. AI can simplify complex contracts, explaining each section in plain English so you know what you're really agreeing to.

☑ **Prep Agreements for the Courts**

Whether you're enforcing or defending a contract, AI can help you draft supporting documents, organize your case points, and get your story on paper *before* it ever hits the courtroom.

Real-Life Example: From Idea to Ink

You're launching a podcast with a homie. Everything's cool until y'all blow up. Sponsors come calling. Merch starts selling. Money gets funny.

Now, one person wants to go solo, and the other one thinks they own the brand.

But there's no contract. No ownership split. No content rights. Just vibes.

Now imagine if, before episode one, you'd used AI to draft a partnership agreement:

- Split the revenue 50/50 unless otherwise agreed
- Define who owns what (logo, episodes, guest list)
- Set rules for exiting the partnership
- Outline what happens if one person stops contributing

That's not paranoia. That's professionalism.

AI lets you *put it in ink*, even when it's just you and your laptop. Because if you're serious about what you're building, then you better protect it like it's already worth millions.

The Hustla's Rule:

"Vibes don't hold up in court. Paperwork does."

Let AI help you build contracts that protect your time, your talent, and your terms. Every boss move starts with an agreement. So write it, read it, and make 'em sign it.

Let's Break It Down: *What AI Can Help You Do Legally*

☑ **Draft Legit Contracts**

We covered this, but here's a reminder: AI can help you create professional contracts that outline clear terms for services, payments, deliverables, deadlines, and exit clauses. Whether you're hiring a freelancer, entering a partnership, or onboarding a new client, AI can set the foundation.

☑ Write a Demand Letter or Legal Complaint

When someone breaks the agreement, you don't have to jump straight to court, but you do need to send something that sets the record straight. AI can help you draft a formal **legal complaint** or **demand letter** that communicates your position in a clear, professional, and enforceable way.

☑ Prepare a Lawsuit to File in Court

If things escalate and you need to take it to court, AI can actually help you **draft the framework of a lawsuit**, including your claims, the facts of your case, and the legal basis for your argument. You'll still need to tailor it to your local court's requirements, but you're walking in with a solid first draft that speaks the language.

☑ Build Your Defense If You're Pro Se

What if *you* get hit with legal action and you don't have a lawyer?

AI can help you understand the complaint, break down legal language, and craft a written response. You can prompt it to help you outline your side, prepare exhibits, and even simulate questions for mediation or court.

That's pro se with power. You're not walking in blind; you've got a blueprint.

Real-Life Example: *A Hustla Protects Their Brand*

You built a brand from scratch. You've got a following. You're selling merch. One day, a bigger company copies your entire aesthetic; colors, slogans, even your ad layout and drops a similar product a week before you do.

You're mad, but you're not helpless.

AI can help you:

- Draft a cease-and-desist letter
- File a DMCA takedown
- Draft a legal complaint outlining IP infringement
- Create a civil complaint to file in small claims or superior court
- And if the heat gets turned back on you, AI can help you structure your defense, prep evidence, and write your rebuttals

All of that used to cost thousands just to get started. Now? You can begin that process for **free**, with a clear prompt and a focused mindset.

Sample Prompt: Drafting a Lawsuit

Persona:
"You are a civil litigation attorney with experience representing small business owners in intellectual property disputes."

Task:
"Draft a civil complaint that outlines a claim of copyright infringement against a larger company that duplicated a brand's social media ad concept and product design."

Context:
"The plaintiff is a Black-owned apparel brand that released a "Protect the Culture" capsule line. A week later, a national retailer launched a nearly identical line using the same slogan and marketing copy. The complaint should state the claim, outline damages, and request injunctive relief."

Critique:
"Avoid overly technical citations. Focus on plain language and include key elements of the claim: who, what, when, how, and why it matters."

Refine:
Ask the AI to generate a simplified version for use in small claims court, and a version formatted with state-level complaint structure for official filing.

The Hustla's Rule:

"If you're gonna build the empire, you better protect the castle."

AI gives you the blueprint to defend your brand, file your claim, or fight your case, even if you're standing in that courtroom alone. Because real CEOs play offense *and* defense.

Stay in Bounds: *Navigating Regulations and Compliance Efficiently*

You can have the dopest product on the block. The cleanest brand. The most loyal customer base. But if your paperwork ain't right with the city, the state, or the feds, your whole operation is at risk.

Let's be real:

Too many of us have started businesses without knowing what licenses we need.

We've posted up at pop-ups with no vendor permit.

We've launched e-commerce sites without privacy policies or terms of service.

We've moved like hustlas, but in the eyes of the law, we were unprotected.

That's not because we're careless. It's because the systems were never designed to make it easy for us.

But now? **We've got tools.**

How AI Helps You Stay Compliant Without Losing Your Mind

☑ **Figure Out What's Required**

You can ask AI:

"What licenses and permits do I need to start a mobile food business in Georgia?"

…and it'll pull together a list, often with links to government sites, specific forms, and cost breakdowns.

☑ **Break Down Legal Jargon**

Have you ever tried to read a state business filing website? It's like decoding hieroglyphics.

AI can translate that mess into plain English:

"This means you need a business license, a seller's permit, and food handling certifications. Here's what each one does."

☑ **Draft Policy Documents**

Selling online? Collecting customer data? Offering coaching or consulting services?

You'll need Terms of Service, Privacy Policies, Refund Policies, AI can help you create all of them, custom to your business.

☑️ Keep You Out of Trouble

AI can't give legal advice, but it can flag issues, give you checklists, and help you stay ahead of compliance before regulators, agencies, or clients ever bring heat.

🐌 Sample Prompt: Compliance Checklist for a Mobile Retail Business

Prompt:

> *You are a business consultant who specializes in small retail startups. What licenses, permits, and compliance requirements are needed to legally operate a mobile clothing boutique in the state of California? Break it down by state, county, and city (Los Angeles). Include any health and safety codes, tax registration, and insurance requirements. Use plain language and give step-by-step instructions.*

The power in this chapter isn't just about writing letters and contracts; it's about **moving right from day one**. You can't afford to build a brand that goes viral just to get hit with a cease-and-desist or a shutdown notice from the city.

The Hustla's Reminder:

"You can't scale chaos."

AI is your compliance co-pilot keeping your business clean, your name solid, and your hustle unstoppable, because when the money starts flowing, you want your paperwork tight and your path clear.

When the Bag Got Blocked: *Real Stories of Paperwork Gone Wrong*

Sometimes, it's not the competition that kills a business, it's the missing paperwork.

Let's look at two real-life situations where **a lack of compliance, protection, or documentation cost people money, ownership, or both**.

These stories aren't here to scare you, they're here to wake you up.

Example 1: TLC – The Best-Selling Group That Went Bankrupt

TLC was one of the biggest girl groups of the '90s. Millions of albums sold. Grammys. Sold-out tours. And yet, **they filed for bankruptcy at the height of their fame**.

Why? Because the contracts they signed early on were trash. They didn't understand the publishing splits, the royalties, or who owned what. And once they blew up, they realized they were locked into an agreement that paid everybody *except* them.

They didn't have the legal knowledge or leverage to negotiate better terms in the beginning, and they paid the price when it mattered most.

Lesson:

Just because the money's coming in doesn't mean it's coming *to you*. Without contracts that protect your IP, your ownership, and your royalties, you could be sitting on gold and still go broke.

Example 2: Slutty Vegan vs. Food Truck Vendors

In 2022, Pinky Cole, founder of Slutty Vegan, made headlines not for her booming business, but because several vendors accused her of not paying invoices and not having clear agreements in place.

One vendor even went public, saying their truck was taken off the road after working with the brand without a formal contract. The back-and-forth was messy, and while Pinky defended herself, **the lack of clear documentation made it harder to resolve quickly and cleanly**.

Now let's be clear, Pinky is still a powerhouse. She's that… But the drama *slowed down the movement*, brought unwanted smoke to the brand, and highlighted a truth we can't ignore:

When the paperwork isn't tight, even the dopest brands can get caught up.

The Hustla's Reminder:

"Talent without structure is a time bomb."

Your business needs paperwork like your body needs a spine. And now? You've got AI to help you stand tall. Don't let a missing contract or overlooked permit be the reason your name ends up in court or on a blog.

Chapter Four

Everyday I'm Hustlin'
Office Automation with AI

You ever feel like your business is running *you*? Emails piling up. Invoices late. Documents unorganized. Scheduling a mess. You're doing everything but getting nowhere.

That's because you're not just the CEO. You're the assistant. The admin. The operations manager. The bookkeeper. The customer service rep. The tech support.

All that and you still expected to grow the business too?

Let's be real: that's not sustainable. That's a setup for burnout, and burnout don't build legacy. But now the game has changed. With AI, you can automate the everyday stuff that eats up your time, and free yourself to focus on the moves that *really* matter.

This chapter is all about how to turn AI into your office assistant; the one that never calls in sick, never forgets a task, and don't need PTO. We're talking automation for scheduling, email replies, file creation, reminders, organization, and more.

Because real hustlas don't get caught up doing busywork.

We automate the routine so we can dominate the vision.

What We'll Cover in This Chapter:

- How to use AI to manage your calendar and appointments
- How to generate reports, letters, and business documents instantly
- Automating follow-up emails, reminders, and workflows
- Organizing your files and creating templates that work for you
- Real prompts to turn AI into your digital assistant. No extra hires needed

The Hustla's Reminder:

"Your grind should move like a system not a scramble."

Automation ain't about being lazy, it's about building lean, smart, and scalable. So, let's get to it and start making your hustle run like a machine.

Time Ain't Free: *Managing Your Calendar and Appointments Like a Boss*

If your time ain't managed, your business ain't moving right.

It don't matter how dope your product is, if you're double-booked, missing meetings, or taking calls that don't push the vision forward, you're leaking energy. And in business, leaked energy equals lost money.

Here's the truth:

A lot of entrepreneurs don't *value* their time, because no one taught us how. We were raised to hustle hard, not to schedule smart. But at this level?

Time is the asset.

AI helps you protect it.

Here's What AI Can Do for Your Calendar Game:

☑ **Set Up and Sync Your Schedule**

You can have AI integrate with Google Calendar, Outlook, or whatever system you use, and then organize your day, week, or month. Just tell it what type of tasks need to be blocked off, and it'll suggest time slots automatically.

☑ **Write Booking Policies and Set Boundaries**

AI can help you craft a professional booking policy, cancellations, late fees, meeting types, all that so clients know the deal before they hit your link. It makes your business look polished and prevents those "Can we reschedule?" headaches.

☑ **Create Automated Responses**

Tired of sending the same "Let me check my calendar" message? AI can draft replies that link to your booking system, confirm appointments, or even suggest alternate times that are ready to send in seconds.

✅ Send Reminders Like a Pro

No-shows will kill your momentum. AI can help you set up automated reminders via email or text, so your clients know when, where, and how to show up.

🦗 Sample Prompt: Manage My Weekly Schedule

You are a virtual assistant. Based on the following list of tasks, create a suggested weekly schedule that blocks time for meetings, marketing, admin tasks, and creative work. Prioritize focused work in the mornings and limit meetings to afternoons. I'm available Monday–Friday from 9am–5pm. Add in two 30-minute breaks and one lunch hour per day.

You can also say:

Draft a booking confirmation email for my coaching sessions. Include meeting link, reschedule policy, and contact info. Keep it friendly but professional.

Real Talk:

When your calendar's running wild, your brand energy feels chaotic. But when your schedule is clean, intentional, and automated? You show up sharp. Focused. In control.

You stop reacting and start *orchestrating*.

The Hustla's Rule:

"You can't scale what you don't schedule."

Let AI run your day so you can run your business. Time is currency; protect it like your bank account.

Plug In the System: *Connecting AI to Office 365, Zoom & More*

Alright, so now you know AI can manage your schedule. But let's take it further and make your AI actually *talk* to your calendar, your inbox, and your Zoom account so you're not just planning smart, you're running smart.

Here's how to plug your AI assistant into your real-life tech stack:

🔌 Microsoft Office 365 Integration

If you're using Microsoft Outlook or the Office 365 suite, you can connect AI platforms (like ChatGPT through Zapier, Microsoft Power Automate, or Copilot) to your account to do things like:

Auto-schedule meetings from emails or form submissions

Draft and send follow-up emails after meetings

Add tasks to your calendar based on prompts you give AI

Organize your inbox by urgency, project, or contact

How to Get Started:

1. Use **Microsoft Copilot** (built into Office 365) for AI-generated summaries, scheduling help, and task creation right inside Outlook or Teams.
2. Or connect AI tools via **Zapier** or **Power Automate** to trigger flows like:
   - "When I get an email with the subject 'Booking,' add it to my Outlook Calendar."
   - "Every Monday, create a new weekly plan in Excel or Planner."

Zoom Integration

Running virtual meetings, webinars, or coaching calls? Zoom can get messy if you're manually doing everything. AI can help:

Auto-schedule meetings and add them to your calendar

Send Zoom links with confirmations instantly

Summarize Zoom meeting notes or generate follow-ups

Remind participants with automated email or text

How to Get Started:

1. Go to your Zoom settings and connect it with your calendar tool (Google or Outlook).
2. Use platforms like **Motion, Clockwise**, or **Notion AI** that can pull your Zoom links into one dashboard.
3. With AI + Zapier:
   - "When a new Zoom meeting is created, send a confirmation email to the guest."
   - "After Zoom meeting ends, auto-draft summary notes and send to my inbox."

Sample Prompt: Set Up My Workflow

You are a digital business assistant. Help me integrate my Microsoft Office 365 account with Zoom. I want to auto-schedule Zoom meetings when someone books through Calendly, send them a confirmation email, and add the meeting to my Outlook calendar. Write this step-by-step so I can follow it easily.

The Hustla's Reminder:

"You don't need a whole team when your tech is in sync."

Integration is automation. Connect your tools, let the AI handle the traffic, and reclaim your time like a boss who knows their worth.

Inbox Boss Moves: *Scheduling, Email, and Virtual Communication*

Let's be real, your inbox might be your business's biggest bottleneck. You're missing opportunities because emails get buried. You're chasing clients who "never saw the invoice." You're spending hours typing the same replies, setting up meetings, and confirming details *when all that could be automated, delegated, or done in seconds with AI.*

In this section, we're talking about **cleaning up your inbox, tightening your communication game, and letting AI do the scheduling dance for you.** This is where your hustle stops being chaotic and starts moving like a well-oiled machine.

Email Management: From Overwhelmed to On Point

Here's What AI Can Do for You:

☑ Write Your Emails for You

Sales emails. Follow-ups. Apologies. Inquiries. Confirmations. AI can handle it all with tone, voice, and precision, just tell it what you want to say and who you're talking to.

☑ Organize and Categorize

Use tools like **Superhuman AI**, **Outlook Copilot**, or **Gmail + Zapier** to categorize your messages: high priority, clients only, invoices, follow-ups, etc. Let the AI filter noise from necessity.

☑ Create Auto-Replies

Sick of typing "Thanks, I'll get back to you soon"? AI can craft auto-responders based on your brand voice; warm, professional, or straight to the point.

☑ Summarize Long Emails Instantly

Stop reading four-paragraph essays. Use AI (ChatGPT, Notion AI, or Outlook Copilot) to summarize long messages in a few bullet points.

📅 Scheduling Like a Grown CEO

When you're doing business, *access to you* should feel intentional, not chaotic.

☑ Sync Your Calendars

Use **Calendly**, **Motion**, or **Google Calendar** integrated with AI to let clients book you without 10 back-and-forths. You can set up buffers between meetings, automatic Zoom link generation, and custom meeting types (e.g., 15-min intro calls vs. 60-min strategy sessions).

☑ Auto-Schedule Based on Priorities

Use tools like **Reclaim.ai** or **Clockwise** to let AI build your weekly schedule based on tasks you feed it. It'll rearrange based on urgency, availability, and your preferred deep work hours.

☑ Set Up Email Triggers That Schedule You

Use **Zapier** to say:

"When someone fills out my intake form or sends a message with the word 'book', send them my Calendly link + availability."

💬 Virtual Communication That Hits Different

☑ AI-Written Meeting Agendas

Tell AI:

Create an agenda for a client strategy meeting focused on marketing and brand expansion.

☑ Generate Follow-Ups Instantly

Prompt AI:

Write a follow-up email thanking the client for the meeting, summarizing what we agreed on, and outlining next steps.

☑ Reply Faster with Smart Suggestions

Tools like **Superhuman AI** or **Gmail Smart Compose** suggest responses based on the email you just received. Customize it to sound like *you*.

☑ Record and Transcribe Virtual Meetings

Using **Otter.ai**, **Fireflies**, or **Fathom**, you can automatically record Zoom calls, generate transcripts, and even ask AI to write follow-up notes, action items, and recaps for the whole team.

🪁 Sample Prompt: Clean My Email Game Up

You are my virtual assistant. I receive 50+ emails a day and important messages get lost. Create a strategy to 1) auto-label client emails, 2) summarize long replies, 3) send follow-up reminders 24 hours after I email someone and don't hear back. Also, write a friendly auto-reply for after-hours emails that lets people know when I'll get back to them.

The Hustla's Rule:

"Every second you spend on email is a second you ain't building the empire."

Let AI manage the inbox, run your calendar, and talk for you so you can spend more time making moves, not typing messages.

Inbox Automation: *Email Ain't the Enemy, It's a System to Be Tamed*

Your inbox should be a revenue tool, not a stress source, but most entrepreneurs are drowning in emails. Missed clients. Forgotten follow-ups. Late invoice reminders. And what's wild? **Most of it can be automated, filtered, or pre-written**, if you know how to connect the right tools and tell AI what to do.

This section walks you through *exactly how to integrate AI into your email game*, so your communication moves like clockwork, even when you're offline.

🪁 Step 1: Choose Your Email Platform + AI Tool

Before anything, you need a setup that plays nice together.

- If you're on **Gmail** → Use tools like **Zapier, Superhuman AI, Gmail Smart Labels**, and **ChatGPT** via browser extension.
- If you're on **Outlook (Office 365)** → Leverage **Microsoft Copilot, Outlook Rules, Power Automate**, and **ChatGPT plug-ins**.
- Bonus Layer: Connect **Zapier** or **Make.com** to build automations between your inbox, calendar, CRM, or task manager.

🧠 Step 2: Create Smart Email Labels or Folders

You want to **teach AI how to separate noise from money.**
Here's how to do it:

Gmail Setup:

1. Go to Settings > Filters and Blocked Addresses > Create a New Filter
2. Filter by sender, subject line (e.g., "Invoice," "Booking"), or keywords
3. Assign Labels like:
    - 🔴 URGENT: Clients Only
    - 💸 Invoices & Payments
    - ✋ Needs Reply
    - 🧠 Ideas / Content Requests

4. Use **Zapier or ChatGPT plug-ins** to watch those folders and auto-summarize, reply, or push tasks to your calendar.

Outlook Setup (Office 365):

1. Go to Rules > Create Rule
2. Set conditions like:
    - If subject includes "payment" or "meeting," move to **High Priority** folder

- o If sender is a client, mark as Important
3. Use **Power Automate** to create flows that say:

"If email is labeled 'Client,' create a follow-up task and send a calendar invite."

Step 3: Draft Auto-Replies That Work While You Rest

Let AI handle that "Thanks for reaching out" vibe while you focus on the next move.

Sample Prompt:

> *You are a professional business assistant. Write a friendly auto-reply email for after-hours messages that lets clients know I'll respond within 24 hours. Include a link to my FAQ and booking page.*

Result:

Hey there   thanks for reaching out! I'm currently offline but will get back to you within 24 hours. In the meantime, feel free to check out my FAQ page or book time with me directly here: [Insert link]. Talk soon!

   Pro Tip: Use different auto-replies for different folders. If it's a new client? Sound welcoming. If it's an invoice request? Be direct and efficient.

Let AI adjust tone based on the situation.

Step 4: Let AI Handle the Follow-Up Game

Most deals are lost in the silence. If you send a pitch, proposal, or quote, and don't follow up within 48–72 hours, you're leaving money on the table.

AI + Email Automation = Closed Loops

Here's how to run it:

Using Zapier + Gmail:

- Trigger: "If email sent from me contains 'proposal' or 'quote' AND no reply is received within 48 hours…"
- Action: "Send follow-up email with a soft nudge + restate the offer."

Prompt to generate follow-up copy:

> *Write a casual follow-up to a potential client I emailed two days ago. Remind them I'm available to answer questions and would love to support their project.*

Step 5: Summarize Long Emails Before You Even Open Them

This is *life-saving* when you're juggling 30+ threads a day.

Use browser extensions like **ChatGPT for Gmail** or Microsoft Copilot to auto-summarize lengthy emails into 2-3 bullet points.

Prompt:

> *Summarize this email thread. Highlight any action items, deadlines, or questions I need to respond to.*

Pro Hustla Moves

- **Link Gmail to Notion or Trello** via Zapier so important emails become to-do tasks
- **Use ChatGPT to bulk-reply to similar emails** (like "Thanks for your interest!" messages)
- **Let AI create weekly digest reports** of your top emails, tasks, and follow-ups

The Hustla's Rule:

"An inbox with no system is just chaos dressed in unread messages."

You don't need an assistant; you just need AI wired to your workflow. Let it filter, reply, follow up, and keep your time sacred. That's how CEOs email.

Pen Game Proper: *Document Creation and Management Made Easy*

Let's talk documents, contracts, proposals, bios, pitch decks, reports, invoices, SOPs, you name it.

As a business owner, your paper trail is more than admin, it's your **proof of hustle**. But the truth is, most of us don't have time to sit down and write everything from scratch every time. And even if we *do* write it, how many times have you lost track of where it's saved, what version is current, or whether it's ready to send?

It ain't that you're disorganized. You're just busy, and nobody ever handed you a system.

But now, AI + cloud tools can turn your document chaos into a smooth, scalable machine.

What AI Can Handle Instantly:

☑ **Draft business documents in seconds**
☑ **Generate polished letters, contracts, bios, and SOPs**
☑ **Create multiple versions for different audiences**
☑ **Help you format, brand, and rewrite like a pro**

1. Create Business Docs in Seconds

Need a one-pager for a client? A capability statement? A welcome packet? A course outline? AI's got you.

Sample Prompts:

> *Draft a one-page service agreement for my event planning business. It should include services offered, timeline, payment terms, and a cancellation policy. Keep it clean and professional.*

> *Write a standard operating procedure for onboarding a new coaching client. Include steps, timelines, communication guidelines, and tools used.*

> *Write a vendor introduction email with a one-paragraph overview of my company and an attached capability statement.*

AI Tip: Want multiple versions? Say:

> *Give me three versions of this document: one formal, one casual, and one culturally relatable for Black small business owners.*

2. Organize Like a Boss with Cloud Storage

Creating documents is only half the battle. The other half? Knowing where everything lives.

Use **Google Drive**, **Dropbox**, or **OneDrive** to build your document HQ.

Se   ke this:

- 💼 Business Essentials
  - o   Contracts
  - ~ Business Licenses
  - Brand Assets
- 📑 Client Files
  - o   Onboarding Docs
  - ^ Proposals
  - nvoices
- 🛠 SOPs + Templates
  - o   Sales Scripts
  - o   Email Templates
  - o   Training Guides

Then name your files with power:

BAD: "invoice final 2 edited.doc"

GOOD: "Invoice_JohnsonEvents_032524_APPROVED.pdf"

3. Integrate AI Into Your Document Workflow

Let's make your docs *move on their own.*

Here's how:

AI + Google Docs (via ChatGPT extension or Workspace add-ons)

- Use prompts like "Summarize this document," "Fix the grammar and tone," or "Rewrite this in a confident voice."
- Let AI edit proposals, client recaps, or internal reports before sending.

AI + Notion / Coda / Airtable

- Auto-generate reports, meeting notes, or weekly summaries from task updates
- Store templates for things like pitch decks, intake forms, or agreements, then use AI to fill them in dynamically

Pro-Level Prompt:

You are a business consultant. Take this bulleted outline and turn it into a two-page pitch deck script for investors. Keep the tone confident and focused. Include a one-line call to action at the end.

4. Auto-Generate Fillable Forms, Docs & More

Use **Tally**, **Typeform**, or **Google Forms** to collect client info

Then use **Zapier** to push that data into auto-filled proposals, contracts, or invoices

Use tools like **Formstack** or **PDF Monkey** to turn data into branded PDFs with **no design skills required**

Imagine this:
- A client fills out a questionnaire
- Their answers trigger a pre-built AI prompt
- You get a completed onboarding document, formatted and ready to sign
- You didn't lift a finger

That's automation at scale.

Sample Prompt: Document Kit Builder

I'm launching a coaching program. I need a complete onboarding doc kit: client welcome letter, coaching agreement, intake form, and FAQ. Make sure each document is in a warm, professional tone and can be reused for future clients.

BONUS: Create a Branded Template Library

Use AI to write once, then store forever.

Build a **Branded Doc Kit** with reusable templates like:

- Service Agreements
- Proposal Cover Letters
- Media Kits
- Pitch Decks
- Invoice Memos
- Coaching Welcome Guides

Store 'em in your cloud drive. When you need one, drop a few details into an AI prompt, and boom, it's fresh, formatted, and fire.

The Hustla's Rule:

"If it ain't documented, it didn't happen."

Put your business in writing. Store it like it matters. Automate it so it never gets lost. That's how grown businesses run. Let AI do the heavy lifting while you focus on the legacy.

Case Study: Time Savings in a Real Business

Let's bring this automation talk out the cloud and into real life.

Meet **Keisha**, a 34-year-old entrepreneur from Atlanta running a boutique branding agency called *Vibe Theory Creative*. Her work is fire; logos, social media makeovers, marketing strategy, but like a lot of us, she was drowning in admin.

Before AI? Keisha's hustle looked like this:
- Waking up to 40+ unread emails
- Spending 2–3 hours daily on back-and-forth scheduling
- Copy-pasting the same client onboarding messages into emails
- Writing contracts and proposals from scratch
- Losing files across Google Drive, Dropbox, and her desktop
- Running her business in a constant state of reactive chaos

Keisha wasn't lazy, she was overloaded.

Then she tapped into AI. Here's how it changed the game:

Problem: Scheduling Nightmares

Solution: She connected Calendly to Google Calendar, synced Zoom, and set up AI-generated autoreplies.

Now: Clients book time directly from her website. They get an automatic confirmation, Zoom link, and prep guide, no manual work.

📭 Problem: Inbox Drain

Solution: Keisha used ChatGPT to write auto-replies and created Gmail filters with Zapier to tag, sort, and prioritize messages.

🔄 Now: VIP client emails are labeled + flagged automatically. She checks her inbox twice a day, not every 10 minutes.

📝 Problem: Proposal Overload

Solution: She built a dynamic proposal template in Google Docs and uses AI to auto-fill key sections based on each client's industry, goals, and scope.

🔄 Now: A process that used to take 45 minutes per proposal takes under 7 minutes, and it's more tailored than ever.

📁 Problem: Lost Docs, Multiple Versions

Solution: Keisha created a branded document library in Notion, using AI to maintain version control and generate new docs from templates.

🔄 Now: All her documents are organized by service line, labeled by client name, and searchable in seconds.

⏱ What Keisha Gained in 30 Days:
- 10+ hours a week saved
- No more missed meetings or late emails
- Happier clients due to faster onboarding
- More mental energy for high-level strategy and creativity
- Room to finally launch her first group coaching offer

Keisha's Takeaway:

"I used to feel like I was running in circles just trying to keep up. Now? AI handles my process so I can focus on my purpose. I don't feel like an employee of my own business anymore; I feel like the CEO."

The Hustla's Reminder:

"Time is a tax the game takes if you don't build systems."

AI won't replace your hustle, but it will return your time, and with time? You don't just grow, you elevate.

Case Study: Scaling Smart – From Chaos to Systems at Legacy Logistics

Meet **Darnell**, a 41-year-old ex-military turned entrepreneur who runs *Legacy Logistics & Freight Solutions* out of Houston. He started the company with one box truck and a dream, and within two years, he was managing five trucks, three contractors, and multiple B2B delivery contracts across the Gulf Coast.

The hustle was real, but so was the operational pressure.

Darnell's Challenges:

- Manual dispatching via text, email, and spreadsheets
- Contractors sending incomplete or inconsistent delivery logs
- Delays in invoicing clients, leading to cash flow gaps
- Late compliance renewals (CDLs, DOT inspections, insurance, etc.)
- No real CRM system, customer info lived in his head or sticky notes

"I was running the business with muscle and memory, but not with systems. I was stuck in the truck instead of building the company."

The AI-Powered Flip

Darnell didn't need more help. He needed automation. So, here's how AI, and a few key integrations, revolutionized *Legacy Logistics*:

1. AI-Enhanced Dispatching System

He connected Google Sheets + Zapier + ChatGPT.

- Contractors submit availability via a Typeform
- Zapier logs it into a shared dispatch sheet
- ChatGPT auto-generates a weekly route plan based on distance, availability, and load priority

- Final plan is emailed to all drivers with delivery instructions written by AI in plain English

🔁 Time saved: 5+ hours per week.

🔁 Mistakes and reassignments dropped by 70%.

2. Invoicing + Compliance Automation

Darnell used **QuickBooks**, **Zapier**, and **ChatGPT** to create a billing workflow:

- Delivery logs from drivers are entered via Jotform
- ChatGPT turns them into clear invoices with client names, rates, and dates
- QuickBooks generates the invoice + auto-sends it to the client
- AI adds a follow-up reminder if the invoice isn't paid in 7 days

Bonus: AI also sends monthly checklists to drivers for compliance (insurance, DOT numbers, CDLs), based on due dates.

🔁 Invoices are now sent within 24 hours instead of 7 days
🔁 Compliance checklist completion went from 50% to 95%

3. CRM + Customer Experience Management

Darnell built a CRM in **Airtable**, with AI as the brain behind it.

- Every new lead from the website triggers a ChatGPT prompt:

"Create a client profile, assign an account rep, and generate a personalized welcome email with a capabilities deck attached."

- AI follows up automatically in 5 days if no response is received.
- For repeat clients, ChatGPT writes tailored project recaps and proposals based on past performance.

🔁 Darnell converted 30% more leads by following up faster
🔁 He went from "I'll send it later" to having proposals out *the same day*

Results in 90 Days:

- Saved 20+ hours a week on admin and dispatching
- Improved client retention and reduced late payments by over 60%
- Cut onboarding time for contractors from 3 days to 1
- Scaled from 5 to 8 active contracts with the *same* team size

"I stopped putting out fires and started building a fireproof business. AI became the manager I couldn't afford but always needed."

The Hustla's Rule:

"If you want to scale, you need systems that don't sleep."

Darnell didn't grow by working harder, he grew by getting *out* of the way.
AI turned his business from a hustle to a machine, and now he's got time to lead, not just survive.

Chapter Five

Big Ad Energy
Speak the Language of the People

Let's get one thing straight:

Ads don't sell. Emotion does.

You could have the dopest product in your city, but if your ads are dry, disconnected, or sound like a robot from 2004, you ain't gonna convert. In today's economy, attention is currency, and **if your words don't stop the scroll, you just got scrolled past.**

But here's the cheat code: **AI can help you craft ad copy that talks to your audience like you've known them since the block.**

We're not just talking about random slogans.

We're talking about ad copy that:

- Sells your story
- Connects with your people
- Builds trust in under 8 seconds
- And moves folks from curious to clicking to copping

Whether you're running Meta ads, Google search campaigns, TikTok promos, or good old-fashioned flyers, this chapter is your playbook for letting AI give your brand **Big Ad Energy.**

What We're About to Cover:
- How to prompt AI to write Facebook, IG, and Google ads that slap
- Using AI to find your voice and speak your audience's language
- Testing multiple versions of ads fast without the burnout
- Personalizing copy for niche offers, cold audiences, and retargeting
- Real prompts to generate hooks, headlines, and CTAs that close

Because look, AI doesn't know your hustle like you do, but once you train it, once you show it your tone, your audience, your goals it becomes the coldest copywriter on your team.

The Hustla's Reminder:

"If your copy don't convert, your content is just noise."

Big Ad Energy ain't about selling, it's about *speaking directly to the bag.* Let's teach AI how to write like you hustle.

Talk That Talk: *Writing Ads That Hit and Stick*

Let's be real, writing ad copy that actually converts is a skill. You ain't just writing words. You're **triggering emotion. Building trust. Creating urgency. Flipping curiosity into clicks.** That's alchemy.

But here's the good news: you don't need to be a copywriting guru to write like one.

AI is your ghostwriter. Your copy coach. Your creative tag-team partner. Once you know how to feed it the right inputs, it can draft Facebook ads, Instagram captions, YouTube hooks, and Google headlines that sound like *you,* sell like *you,* and speak *your people's language,* because here's the truth: **if your ad don't feel personal, it gets ignored.**

The Anatomy of a Killer Ad Prompt

Every ad needs three things:

1. **A Hook** – Grabs attention
2. **A Value Hit** – Solves a problem or speaks to a desire
3. **A CTA (Call to Action)** – Tells 'em exactly what to do next

Now here's how to get AI to deliver all three like a seasoned street team:

🪶 Sample Prompt: Sell a T-Shirt That Tells a Story

You are a culturally fluent copywriter who creates bold, authentic ad copy for Black-owned fashion brands. Write three Instagram captions (under 125 words) to promote a Juneteenth t-shirt drop. Focus on legacy, pride, and cultural ownership. Each version should have a strong emotional hook, a clear value prop, and a call to action that fits the tone (not "Buy now" make it feel like a movement).

☑ What AI Gives You:

- Caption 1: Legacy-driven, poetic intro, CTA = "Rock the message. Wear the mission."
- Caption 2: Hype energy, slang-heavy, CTA = "Cop yours before they ghost."
- Caption 3: Rooted in history, uplifting tone, CTA = "It's more than merch. It's memory."

📢 What Makes Copy *Persuasive* (And How AI Nails It)

✔ It Speaks to One Person, Not the Whole Room

You want your copy to feel like a DM, not a press release. AI can match tone based on audience description.

Prompt: *"Write this ad for a 28-year-old Black mompreneur juggling her kids and her candle business."*

✔ It Solves a Specific Pain Point

Don't just hype the product, hit a nerve.

Prompt: *"Write an ad that speaks to new business owners overwhelmed by social media. Offer a solution that feels like relief, not pressure."*

✔ It Feels Like It's From *You*

Your audience needs to recognize your vibe. Use AI to match your slang, tone, and voice.

Prompt: *"Rewrite this ad in a tone that blends Nipsey Hussle's mission-driven energy with Issa Rae's relatable confidence."*

✔ It Tells a Story or Paints a Picture

Prompt: *"Write a Facebook ad that tells a 3-line story about starting a business with $50 and turning it into a brand."*

⚠ Don't Just Copy → Coach It

AI is smart, but it's not you. So coach it like it's your intern:

- "Make it sound more hype."
- "Add some rhythm to the flow."
- "Use cultural language that reflects the South."
- "Try 3 versions; one poetic, one bold, one short + punchy."
- "Now make it a caption for a reel under 90 characters."

You're not just prompting. You're *directing*.

The Hustla's Rule:

"You don't need a million-dollar ad budget if your words hit like a movement."

AI can't sell your dream, but it can write like it believes in it. Teach it your tone. Feed it your fire. And let your copy do the closing.

Run the Play: *AI for Ad Strategy & Audience Lock-In*

Having good copy is dope. But if you're aiming it at the wrong audience? You're just throwing darts in the dark hoping to hit something.

This ain't the spray-and-pray era.

This is the precision era.

You don't just need ads. You need strategy. And that starts with knowing exactly who you're talking to, where to find them, and how to hit their mindset.

The game changed. And AI can now do in minutes what used to take teams of analysts, media buyers, and marketing strategists.

Whether you're running ads for a brand, a digital product, or a physical hustle, this section shows you how to turn AI into your **marketing mind reader**.

🧠 Step 1: Define the Target with Real-World Flavor

Before AI can write a fire ad, it needs to know **who's supposed to feel it.** This ain't *"18-35, lives in the U.S."* nonsense. That's lazy targeting.

Tell AI the *soul* of your audience:

- Their lifestyle
- Their daily frustrations
- What they've tried and failed at
- What inspires them
- What they call success

Sample Prompt:

> *You are a Black marketing strategist who creates ad strategy for mission-driven fashion brands. Break down the audience for a streetwear label rooted in HBCU culture. Include psychographics, motivators, cultural references, and platforms they use.*

☑️ What AI gives you:

- "Values legacy, representation, and cultural pride"
- "Spends time on Instagram, TikTok, and in sneakerhead Discords"
- "Distrusts mass market brands but loyal to niche creators with a cause"
- "Buys limited drops and reposts brands that speak their truth"

Now we're not guessing. We're building **ads with aim.**

🎯 Step 2: Build the Ad Strategy Like a Campaign, Not a Coin Toss

AI can help you map the full journey:

- **Awareness Ads** – Who are you, why should they care?
- **Engagement Ads** – Hit their inbox, drop some value, ask a question.
- **Conversion Ads** – Push the offer when the trust is built.
- **Retargeting Ads** – Hit the ones who clicked but didn't cop.

Prompt Example:

> *Build a 3-phase Meta ad campaign for a Black-owned wellness brand launching a new herbal tea line. Include awareness, engagement, and conversion strategies with messaging ideas and platform recommendations.*

AI can also break down:

- Budget tiers ($100 vs $1,000/month)
- Timing and frequency
- A/B test copy vs creative
- What ad styles hit on which platforms (Reels vs static vs Stories)

•• Step 3: Use AI to Spy Without the Cease-and-Desist

Want to know how your competitors moving? Ask AI.

> *Give me a breakdown of the type of ads competitors like The Lip Bar, Bevel, and Actively Black are running. What audience are they targeting, what language are they using, and how are they positioning their brand?*

AI can't pull live ad libraries, but it can analyze trends, synthesize reviews, and identify brand voice based on publicly available content. Combine that with real Facebook Ad Library research, and you're a step ahead.

Bonus Tip: Build an Audience Persona Playbook

Prompt:

> *Create a target audience profile for my AI-powered coaching service for minority entrepreneurs. Include name, backstory, career struggles, goals, fears, and the kind of ad language that would hook them emotionally.*

☑ Result:
- "Meet Danielle, 34, single mom in Philly, running a T-shirt brand and side-hustling as a notary…"
- AI then spits ad ideas, content angles, and platform priorities

The Hustla's Rule:

"If you ain't speaking your audience's language, you're just another ad in the scroll."

Let AI break down the who, the where, and the why so every dollar you spend feels like it was built for impact. Run the play with purpose.

Ad Ops Like a Boss: *Managing Campaigns Without Losing Your Mind*

Look, posting an ad is easy.

Managing it? That's the part that separates the amateurs from the ones who really see returns.

Boosting a post and hoping for the best? That's not strategy, that's donation.

If you want results, you need to track, tweak, and test like a real CEO.

But here's the thing, they made ad management sound like rocket science on purpose.

All those dashboards, split tests, pixels, CPCs, ROAS? It's enough to make you wanna just post and pray.

But AI flips the script.

AI helps you manage your ad campaigns like you got a digital marketing agency in your back pocket, minus the monthly retainer.

🧠 How AI Helps You Manage Ad Campaigns Like a Real One

☑️ 1. Monitor Campaign Performance in Plain English

You don't need to stare at charts all day.

Prompt:

> *You are my digital marketing analyst. Based on the ad campaign data below (copy/paste performance summary), explain what's working, what's underperforming, and give me 3 recommendations to improve click-through rate.*

Let AI break it down like:

"Your image ads are performing better than Reels, shift more budget there."
"Tuesday–Thursday sees the best engagement. Consider pausing weekend ads."
"Your headline isn't hitting, test one that uses urgency."

✅ 2. Write and A/B Test New Variations Fast

You don't need to write 10 different ads from scratch. Let AI generate versions with different hooks, angles, and tones.

Prompt:

> *Rewrite this Instagram ad in 3 different styles: one emotional, one funny, and one with a direct CTA. Keep the product focus the same but make each one feel unique.*

Now you've got material to test without burning time or brain cells.

✅ 3. Generate Reports for You or Your Clients

Whether you're managing your own ads or running them for clients, AI can auto-generate campaign summaries that make you look like a data god.

Prompt:

> *Summarize this week's Facebook ad performance for my skincare brand. Include cost per lead, top performing creative, audience insights, and a recommendation section. Write it like a weekly client update.*

Now you've got a fire report you can send out, post in Slack, or use to make real-time decisions.

✅ 4. Automate the Boring Parts with Zapier & Tools

1. 🛠 Tools to connect:
2. Zapier
3. Meta Ads Manager
4. Google Sheets
5. Slack / Notion / ClickUp

Use it to:

1. Send a Slack alert when cost per click rises too high

2. Auto-tag top performing ads for weekly review
3. Pause low-performing ads automatically after 3 days
4. Log all lead info into a CRM or Notion doc without touching it

Pro Hustla Flow

Your Weekly Ad Workflow with AI

Monday: Ask AI to summarize last week's ad results
Tuesday: Use AI to write 3 new creatives and headlines to test
Wednesday: Plug those into your ad platform and launch variations
Thursday: Have AI create a performance brief with adjustments
Friday: Drop that insight into your CRM, team chat, or Google Doc

That's how you build consistency. That's how you manage at scale. That's ad ops with no anxiety.

The Hustla's Rule:

"Ad management ain't sexy, but profit is."

Stop posting and hoping. Start managing and multiplying. With AI, your campaigns move like they got a CMO behind 'em because now, they do.

Setup 1: *Meta Ads + ChatGPT + Google Sheets*

This workflow lets you track, analyze, and optimize your ad campaigns using AI + Google Sheets with no analyst degree required.

Step-by-Step: Meta Ads Campaign Tracking with ChatGPT + Google Sheets

Step 1: Export Your Ad Data from Meta
- Go to Meta Ads Manager
- Click the campaign you want to track
- Click Reports > Export Table Data > CSV
- Open the file in Google Sheets

Step 2: Clean Up Your Data Sheet

- **Keep key columns like:**
  - Ad Name

- o Click-Through Rate (CTR)
- o Cost per Click (CPC)
- o Impressions
- o Amount Spent
- o Conversions (or Leads)

Step 3: Copy/Paste That Data Into a Prompt for ChatGPT

Prompt:

> *You are a paid ads expert. I'm running Facebook ads for my Black-owned meal prep company. Based on this data [paste table or summarize], tell me:*
>
> 1. *Which ads are underperforming*
> 2. *Which ones I should scale*
> 3. *3 ways to lower my CPC*
> 4. *Which creatives need to be retired*

☑ **Result: A quick breakdown in plain English, including suggestions for what to test, stop, or scale.**

Step 4: Use AI to Write New Ad Variants Immediately

Prompt:

> *Rewrite this headline to test 3 new hooks. Make one emotional, one funny, one focused on urgency.*

Setup 2: Zapier + Slack Alerts for Campaign Monitoring

Let the AI Tap You When It's Time to Adjust

This flow lets you set up a system where you're alerted *only when something needs attention*. Let Zapier act like your AI watchdog.

Step-by-Step: Setup Smart Ad Alerts Using Zapier

Step 1: Connect Meta Ads Manager to Google Sheets
- Use Zapier to log campaign performance to a running sheet
- Zap Trigger: "New ad performance data from Meta"
- Zap Action: "Update Google Sheet with latest metrics"

Step 2: Add Conditional Logic for Slack or Email Alerts

- Zap Trigger: "If CPC > $1.50 OR CTR < 1.0%"
- Zap Action: "Send message to Slack (or Gmail) that says:

  " *Yo! Campaign X is underperforming. CPC just hit $1.87. Might be time to pause or rework.*"

Bonus: Customize messages by campaign name or audience segment so you know exactly what to tweak.

Pro Hustla Bonus Prompt:

You're my AI ad strategist. Based on the last 7 days of my campaign (data below), give me a play-by-play recommendation list:

- *Ads to pause*
- *Creatives to update*
- *Audiences to retarget*

Then write a 3-line performance update I can send to my client or investor.

The Hustla's Rule:

"If your campaign ain't talking to you, it's wasting your money."

Set the tech up once and let it watch the numbers while you work the vision. With AI and the right integrations, your ad game moves like it's funded, even if you're still flipping that first $500.

Flip the Budget: *Optimizing Ad Spend for Real ROI*

Let's get something straight:

Throwing money at ads without a plan is just burning the bag in public.

Whether you've got $50 or $5,000 to spend, **your ad dollars need to move like soldiers;** strategic, targeted, and ready to bring something back.

That's what we call **flipping ad spend into ROI** (Return on Investment), and guess what? AI can help you do it like a seasoned media buyer.

You're not just "boosting posts," you're building campaigns that stretch your budget, scale what works, and cut what doesn't *before* it bleeds.

🏵 Step 1: Set Smart Budget Goals

Before you even launch an ad, AI can help you define what success looks like.

Prompt:

> *You are a digital ad strategist. I'm running Facebook and IG ads for my handmade jewelry brand. My budget is $300 for the next 14 days. Help me decide how to split the budget across brand awareness, engagement, and conversions. Also recommend a benchmark for what a good cost per purchase should look like.*

☑ What AI gives you:

- Budget split (e.g. 40% awareness, 30% engagement, 30% retargeting)
- Daily cap to avoid overspend
- Benchmark metrics to track (CPM, CPC, ROAS)

Step 2: Track the Right Metrics, Not Just "Likes"

You ain't optimizing for vibes. You're optimizing for **return**.

Let AI help you analyze:

- **CPC (Cost per Click)** – How much are you paying for attention?
- **CTR (Click-Through Rate)** – Is your copy converting curiosity into action?
- **CPA (Cost per Acquisition)** – What's it costing you to get a sale or lead?
- **ROAS (Return on Ad Spend)** – For every $1 in, how much comes out?

Prompt:

> *Review this week's ad data (copy/paste) and tell me which campaign has the highest ROI, which one should be paused, and how I can improve the ad that's costing me the most per lead.*

Step 3: Run AI-Powered A/B Tests Without the Guesswork

Let's say one ad is performing, but you're not sure why. Or maybe it's close but not quite hitting.

Use AI to test:

- Headlines
- Visuals
- Offers
- CTAs
- Audiences

Prompt:

> *Rewrite this offer in 3 different ways: one with urgency, one with humor, and one based on a customer pain point. Keep it short and punchy and under 15 words each.*

Then run each version with the same image. Watch what wins.

Pro Tip: Use Google Sheets or Notion to log every test. Let AI summarize what worked and what didn't each week.

Step 4: Scale the Winners. Kill the Dead Weight.

Every ad campaign teaches you something. But the only ones that matter are the ones that bring money back in.

Prompt:

> *Based on this 14-day campaign data, create a simple report that shows me what to scale, what to cut, and how to reallocate my $150 left for the month to get the most sales.*

AI will look at performance patterns and recommend:

- What ad to double the budget on
- What audience is converting best
- What platform deserves more love (FB vs IG vs TikTok)

Bonus Hustla Flow: Budget Tracker System

Set up a weekly AI-powered budget review.

1. **Monday** – Copy/paste last week's spend and performance data

2. **Prompt AI** to summarize what gave you the highest ROI
3. **Wednesday** – Ask for 2 new creative angles to test
4. **Friday** – AI writes you a 3-sentence "investor-ready" update with your current return stats

The Hustla's Rule:

"Every dollar should clock in and clock out with purpose."

AI don't just make ads easier; it makes them smarter. You don't need a million-dollar budget. You need a plan, a prompt, and the mindset to flip every click into capital.

Chapter Six

Mo' Marketing Mo' Money
Spend Right Message Tight

Here's the truth they don't teach you in most business books, *"Marketing ain't just about getting attention, it's about keeping it."* You can have the dopest product, the cleanest visuals, and the flyest logo…

If you're not marketing with *strategy*, consistency, and emotion? You're just talking to yourself. Marketing is how you build **trust at scale**. It's how you stay top of mind when your audience isn't ready to buy yet, and how you make sure they circle back when they are, and right now? **AI is the ultimate marketing partner.**

This chapter isn't about going viral. It's about building a brand that lasts. A strategy that moves every day without you burning out, breaking the bank, or begging for engagement.

What You're About to Learn:
- How to build a full marketing strategy with AI, from idea to execution
- Using AI to repurpose content across platforms (work smart, not twice)
- Automating email campaigns, funnels, and brand messaging
- Marketing for the long haul; not just the algorithm
- Real-world prompts to map your content calendar, build campaigns, and hit new markets

Because **mo' marketing done right = mo' money,** and with AI? You don't need a marketing team; you just need the vision and the prompts.

The Hustla's Reminder:

"Visibility creates viability."

The more people know your name, your value, and your why, the easier it is to turn attention into income.

Let's build a marketing machine that moves like you do; consistent, creative, and unapologetically locked in.

Read the Room: *Market Research & Audience Insight with AI*

The first rule of marketing? **Don't guess, know.**

You can't speak to your audience if you don't understand what they want, what they need, and what they're tired of hearing. Too many businesses are shouting into the void, saying what they *want* to say instead of what their people actually need to hear, but here's where AI changes the game.

With the right prompts, AI becomes your market research assistant pulling trends, identifying gaps, and helping you *see your customer clearly* before you spend a single dollar on ads, content, or email blasts.

No surveys. No guesswork. No data scientists on payroll.

Just targeted insight served hot and fast.

🔍 Step 1: Define the Market Like a Mind Reader

You don't just want demographics, you want *psychographics*. What motivates them? What language do they use? What keeps them up at night?

Prompt:

> *You are a brand strategist. Break down the target audience for a natural skincare line for Black men. Include age range, mindset, buying habits, pain points, platforms they use, and what type of content grabs their attention.*

☑ AI Gives You:

- "Men aged 25–45, conscious about self-care but underserved by traditional skincare brands."
- "Wants simplicity, transparency, and products that feel masculine without being toxic."
- "Scrolls IG, watches grooming tutorials on YouTube, listens to podcasts about culture + wellness."
- "Trusts brands with community ties and real representation."

Now you're not just talking *to* them, you're talking *for* them.

🫐 Step 2: Spy the Competition (Legally, of Course)

You don't need to reinvent the wheel; you just need to make it spin in your direction.

Prompt:

> *Compare the brand positioning, tone, and content strategy of these three businesses: The Lip Bar, Topicals, and Fenty Skin. What audiences are they targeting, and how can a new skincare brand differentiate?*

Let AI study their websites, social posts, and product messaging to find **what they're doing and what they're *not* doing.**

This is how you find your lane in a crowded market.

Step 3: Pull Industry Trends on Demand

You don't need to spend hours Googling what's hot.
Ask AI for a quick hit of industry intel.

Prompt:

> *What are the top marketing trends in the wellness space for Q1 2025? Include consumer behavior shifts, platform trends, and emerging product categories.*

AI Might Say:
- "Demand for mental health-focused products is rising."
- "Instagram engagement is down, but TikTok Lives and Twitter threads are gaining traction."
- "Buyers are gravitating toward brands with personality, transparency, and educational content."

Now your moves are backed by **real-time insight, not vibes.**

Bonus Prompt: Build a Full Audience Profile

> *Create a target audience persona for a business funding consultant who serves women of color entrepreneurs. Include: name, age, city, current struggle, business type, social media use, content preferences, and buying mindset.*

☑ Result:

"Meet Jasmine. 36. Lives in Charlotte. Runs a boutique PR firm. Wears 12 hats. She's tired of Googling grants and piecing together advice from IG Lives. She wants someone who knows the game and can give her a real plan. She watches short-form video and reads punchy, no-fluff newsletters."

You now know exactly how to speak to her.

The Hustla's Rule:

"You can't sell to who you don't understand."

Market research ain't optional, it's the map. Let AI hand you the compass so every word, ad, and offer hits like it was made just for them.

The Master Plan: *Building a Marketing Strategy with AI*

Now that you know who you're talking to, it's time to plan how you talk to 'em.

This is where most people fumble.

They either do too much and burn out, or they do too little and disappear. Inconsistent marketing doesn't build trust. It builds confusion, and confused audiences don't buy. That's why you need a **strategy that's consistent, intentional, and automated** and you don't need a team of marketers, you just need the right AI prompts.

🪨 Step 1: Map Out the Customer Journey

Your customer has stages, awareness, interest, trust, action. AI can help you create content that meets them **where they are**, not where you wish they were.

Prompt:

> *You are a marketing strategist. Map out a content strategy for a fitness coach targeting millennial Black women. Include awareness, engagement, and conversion phases. Show what type of content works best at each stage.*

☑ AI Might Say:

- **Awareness:** "Short-form reels showing transformation stories, funny wellness memes, 15-sec workout tips"

- **Engagement:** "Behind-the-scenes, Instagram Lives, interactive polls, blog posts about emotional eating"
- **Conversion:** "Client testimonials, limited-time offers, DM scripts, and email sequences that push urgency"

Step 2: Build a 30-Day Marketing Calendar

This is how you stay consistent without burning out. Let AI build your whole month based on your audience, platform, and brand tone.

Prompt:

> *Create a 30-day marketing calendar for a Black-owned streetwear brand launching a new collection. Include daily Instagram post ideas, story prompts, email subject lines, and one live event idea.*

Result:
- Week 1: Hype the collection story
- Week 2: Tease designs + BTS content
- Week 3: Launch countdown + influencer shoutouts
- Week 4: Showcase customer fits + drop reviews

Now your content *flows*. No scrambling. No guessing.

Step 3: Plug Into Real-Time Trends

Let AI scan the streets and social media to keep you fresh.

Prompt:

> *What are 3 trending hashtags or content themes in the natural hair space this month? Give me ideas on how to incorporate them into my brand voice without looking like I'm copying.*

AI Might Suggest:

- "#TexturedTuesday is trending. Create a product spotlight with customer stories"
- "Use the 'POV' trend on TikTok to tell your brand's origin story in 10 seconds"
- "Create a 'Things I Wish I Knew About My Hair' carousel for engagement"

Step 4: Automate the Execution

Plan is nothing without the *push.*

Connect your strategy to tools like:
- **Notion** (for planning)
- **Later or Metricool** (for auto-posting)
- **ChatGPT or Jasper** (for writing the posts)
- **Canva** (for designing the visuals)

Prompt:

> *Turn this 30-day plan into a social media post tracker with columns for postdate, platform, caption, visual style, and CTA. Format it for Google Sheets or Notion.*

Now you're not just creating, you're **running systems.**

Bonus Hustla Prompt: Quarterly Campaign Builder

> *Build a 90-day marketing campaign for a startup haircare line. Break it into 3 monthly themes. Each month should have a signature offer, 2 email campaigns, content ideas, and 1 virtual event to drive engagement.*

This one prompt gives you an entire *quarter* of consistent visibility, value, and conversion, all laid out like a blueprint.

The Hustla's Rule:

"A plan ain't real until it's scheduled."

You can't market when you're tired, distracted, or reactive. Let AI carry the planning load so you can show up like the brand you're building.
Strategic. Intentional. Unstoppable.

Inbox Gold
Building Email Campaigns That Convert

They said email is dead. But guess what? **Email still brings in more ROI than almost any platform out.**

Why? Because once they're on your list, you're not fighting algorithms, you're talking direct.

But here's the problem:

Most folks treat email like an afterthought. Dry subject lines. Long-winded paragraphs. No strategy and then wonder why nobody's clicking.

AI is here to change that.

This ain't about spamming inboxes, it's about **building trust, delivering value, and calling your people to take action.**

✉ Step 1: Build a 5-Part Email Campaign with AI

Let's say you're launching a new service or product. AI can write a full campaign from scratch including subject lines, preview text, and CTAs.

Prompt:

> *You are an email copywriter who writes high-converting campaigns for mission-driven brands. Write a 5-email sequence for a new online course helping Black women entrepreneurs build business credit. Include emotional storytelling, pain points, solutions, and urgency. Keep each email under 300 words.*

☑ AI Might Deliver:

- Email 1: "Why no one ever taught you the money game"
- Email 2: "The 3 myths keeping your biz from funding"
- Email 3: "Proof it works. Real stories from the community"
- Email 4: "Doors close soon. Let's build this legacy"
- Email 5: "Final call! Don't let this opportunity pass you by"

Each email is a step in the buyer journey, *awareness, trust, action.*

Step 2: Create Automated Email Funnels

Pair your AI-written emails with email platforms like:

- **MailerLite**
- **ConvertKit**
- **Flodesk**
- **ActiveCampaign**

Bonus: AI can also help you **segment your audience** and tailor messages to different groups.

Prompt:

> *Write two versions of the same email, one for new subscribers and one for people who've opened the last 3 emails. Keep the tone consistent but change the CTA based on readiness to buy.*

Now you're not just emailing, you're **emailing with precision.**

Step 3: Let AI Write Your Newsletters Weekly

Consistency is king, but it can get exhausting.

Prompt:

> *Based on this week's industry trends [paste summary], write a 250-word newsletter for my followers that shares 1 insight, 1 resource, and 1 action step they can take today.*
>
> *Tone: real, relatable, motivating.*

Drop that into your email platform, hit schedule, and keep it moving.

Want it formatted?

Ask: "Format this for MailerLite with subject line, preview text, and bolded key lines."

Step 4: Make Sales Without Sounding Like a Salesperson

Nobody likes being sold to, but everyone loves to feel seen. Let AI help you write sales emails that feel like **conversations, not commercials.**

Prompt:

> *Write a warm, culturally aware sales email for my consulting program. Focus on how I help overwhelmed entrepreneurs simplify their systems and get their time back. Include a soft CTA that invites them to book a free call.*

Bonus Hustla Flow: Weekly Email System
1. **Monday:** Ask AI to generate your newsletter draft
2. **Tuesday:** Use AI to tweak subject lines (test 3 variants)
3. **Wednesday:** Ask AI to write a segmented follow-up email
4. **Friday:** Let AI summarize email performance and suggest improvements

The Hustla's Rule:

If you can get in their inbox, you can get in their bag.

Email ain't about volume, it's about value.

Let AI handle the writing so you can focus on connecting. One click at a time.

Funnels on Autopilot
AI Systems That Sell So You Don't Have To

Let's be real, most of us never heard the word "funnel" until we got deep in the entrepreneur game.

And even then, it sounded like some techy guru talk.

But once you understand what a funnel really is?

It changes how you move.

Because when done right, a funnel becomes your **24/7 salesperson**, your **digital assistant**, your **customer journey on autopilot.**

What Is a Funnel (For Real)?

A funnel is simply this:

A step-by-step path that guides someone from not knowing you to trusting you and then buying from you.

That's it.

Think of it like a first date that turns into a relationship:

1. **They see you** (an IG post, ad, or live)
2. **They get curious** (click your link, download a free guide, or sign up for a workshop)
3. **They get value** (through emails, content, or messaging that helps them solve a problem)
4. **They get the offer** (a paid service, product, course, or consultation)
5. **They buy** (because you built trust *before* asking for money)

That's a funnel.

Not some scam. Not some hack.

Just a process. A system. A strategy.

And with AI? You don't need to hire a strategist, build everything from scratch, or figure it out over months of trial and error.

You just need the vision and the prompts.

Step 1: Let AI Design Your Funnel Blueprint

You don't need to guess what steps to build. Ask AI to map it for you based on your business type, audience, and offer.

Prompt:

> *You are a marketing funnel expert. Map out a customer journey funnel for a business funding coach who helps Black women secure startup capital. Break it into steps from awareness to conversion. Include free content, lead magnet ideas, email touchpoints, and the final offer delivery.*

AI might say:

- **Awareness**: Instagram Reels + "How I Got $10K in Grant Funding" posts
- **Interest**: Free PDF – "Top 10 Funding Resources for New Entrepreneurs"
- **Trust Building**: 5-part email series educating them on how funding works
- **Conversion**: 60-minute webinar + CTA to book a funding strategy call

That right there is a funnel.

It turns attention into action step by step.

Step 2: Create Every Funnel Piece With AI

Once the blueprint is clear, AI can help write the actual content:

The Lead Magnet – Your freebie, workshop, checklist, or guide

Landing Page Copy – Where people sign up

Email Sequence – What they get after opting in

Offer Messaging – The part that closes the sale

Sample Prompt:

> *Write a 3-email sequence for someone who downloaded my "Business Credit Starter Kit." The goal is to educate, build trust, and invite them to book a 1-on-1 consult. Make it warm, empowering, and action-oriented.*

Step 3: Automate the Journey

Once everything's written, you set the system to run *on its own.*

Here's the tech flow:
1. **Connect your opt-in form (Mailchimp, ConvertKit, Stan Store, Flodesk)**
2. **Trigger the AI-written email sequence** once someone signs up
3. **Use Zapier or your email platform** to send reminders, content, and nudges
4. **Insert booking links or purchase buttons** at the right time when trust has been built

You've now got a system that brings people into your world, warms them up, and makes the sale *while you sleep.*

Real-World Example: Funnel for a T-Shirt Brand

Business: Culture-based streetwear brand
Audience: Black Gen Z college students
Funnel Flow:
- **Step 1 (Awareness):** IG Ad that says, "Who Else Is Tired of Seeing Us Copied and Not Credited?"
- **Step 2 (Freebie):** Free wallpaper or phone lock screen that says "Protect Black Culture"
- **Step 3 (Email Sequence):** Story behind the brand, creator's journey, upcoming drops, early access
- **Step 4 (Offer):** "VIP Access to Our Next Drop + 15% Off for Subscribers Only"

That's culture. That's connection.

That's automated marketing built for *your* audience.

Bonus Prompt: Funnel Flow Builder

I help parents of children with special needs navigate the education system. Build me a complete funnel that offers a free "504 & IEP Prep Kit," educates them on their rights, and leads them to book a strategy session. Include landing page copy, 4 emails, and 2 content ideas to post while the funnel runs.

The Hustla's Rule:

"If your business only works when you're awake, you ain't scaling."

Funnels let you multiply your message. AI lets you build them faster, smarter, and with zero guesswork. Set it up once, and let it talk for you, teach for you, and sell for you. On autopilot. Every day.

Chapter Seven

It Was All a Stream
AI in the Social Media Game

Social media used to be simple. You posted a pic, added a caption, maybe hit a few hashtags, and boom you had motion.

But now?

It's an algorithmic jungle.

Every platform got its own language. Your people are scrolling fast. Trends change by the hour. And if you're not showing up consistently with content that connects, you disappear.

But here's the thing:

You don't need to go viral. You just need to stay valuable.

This chapter is about how to use AI to **run your social media like a content machine** without selling your soul, losing your time, or burning out trying to keep up with TikTok dances you ain't built for.

What You'll Learn in This Chapter:

- How to use AI to plan, write, and schedule social content across platforms
- How to stay on-brand, culturally relevant, and consistent, even on a budget
- How to let AI help you remix your content into Reels, threads, carousels, and captions
- Tools and prompts to generate content calendars, scripts, and short-form bangers
- How to measure and adjust based on what's really working and not just likes

Because social media shouldn't feel like a second job. It should feel like an extension of your message, your brand, and your mission.

The Hustla's Reminder:

"If content is king, consistency is the throne."

Let AI help you stay visible, stay relevant, and stay on brand so you can build presence without losing your peace.

Post Up with Purpose
Content Creation & Scheduling with AI

Let's keep it a buck, content is currency, but if you're waking up every day wondering, "What should I post today?" you're already behind.

Consistency is what builds trust, and trust is what converts followers into **clients, customers, and community.**

But most of us don't have time to sit around writing captions, editing videos, and brainstorming reels. That's a full-time job by itself.

That's why this section is about **letting AI become your content team**, your writer, your scheduler, and your strategist. All gas, no burnout.

Step 1: Use AI to Plan Your Content Calendar in Minutes

Prompt:

> *You are a social media content strategist for a Black-owned coffee brand. Create a 30-day content calendar for Instagram that includes post types (reels, carousels, memes), captions, themes (legacy, hustle, culture), and CTA ideas. Audience: Young professionals who care about quality, culture, and community.*

AI Gives You:
- Day 3: "#CaffeineAndCulture" carousel – Highlight Black artists in your city
- Day 12: Reel – Behind-the-scenes of the roasting process
- Day 18: "What fuels your hustle?" – Engagement post with poll
- Day 26: Testimonial spotlight + 10% off CTA

You just got a whole month of content with vibe and structure.

Step 2: Let AI Write Your Captions & Scripts

Tired of caption writer's block? Let AI cook.

Prompt:

> *Write a caption for a Black woman-owned skincare brand dropping a new serum. Make it warm, empowering, and sound like something Issa Rae would post.*

Result:

For skin that glows through grind mode. Our new "Balance Serum" just dropped, and it's built for your 5am meetings, your late-night grind, and every selfie in between. Glow up on your own terms.   #BuiltByUs

Want a script for a 15-sec Reel?

Write a quick, punchy script for a Reel promoting the same serum using a transformation format. Make it catchy, confident, and relatable.

Step 3: Schedule Everything with Automation Tools

Now that AI's written it, get it **posted without lifting a finger.**

Tools like:
- **Metricool**
- **Lately**
- **Buffer**
- **Planoly**

Let you:
- Schedule content across platforms (IG, TikTok, Twitter, YouTube Shorts)
- Add AI-written captions + hashtags
- See best posting times based on engagement
- Preview how your feed will look before it drops

Pro Tip: Create a "Content Drop Day" once a week. Use AI to write, revise, and load up your next 7 days of content. Then you're done.

Bonus Flow: Repurpose Like a Pro

AI can turn one piece of content into five.

Prompt:

Repurpose this IG caption into 1 tweet, 1 email teaser, 1 story script, and 1 blog intro. Tone: motivational, direct, and still rooted in culture.

Now you're everywhere, and your message is *moving*.

The Hustla's Rule:

"You don't need to post every day; you just need a system that shows up for you."

AI gives you your time back, your voice on repeat, and your content in motion.

Because content isn't king, **consistency with clarity is.**

Post Up With Purpose: Content Creation & Scheduling with AI

Social media don't stop, but you need to.

Because if you're waking up every day trying to post in real time? You're hustling backwards.

That "What should I post today?" stress is the quickest way to fall off.

You need a system. A rhythm. A strategy that frees your brain and feeds your brand.

This section is the blueprint.

We're showing you how to turn one day like a Sunday, into your content batching session for the whole week.

And we're putting AI to work, so you never run out of ideas or energy again.

Step 1: Use AI to Plan Your Week in One Sitting

Sunday is your strategy day.

That's when you open your laptop, grab your content vault (blog post, video, podcast, product page, etc.), and let AI go to work.

Enter: Lately.ai – Your AI Content Machine

Lately.ai is like your ghostwriter, strategist, and scheduler *all in one*. It takes long-form content (blogs, podcasts, videos, YouTube links, or even your own social pages)

→ splits it into bite-sized social media posts

→ and then lets you **schedule everything right from inside the app.**

No jumping to Hootsuite, Buffer, or Later.

It's all built-in.

Real Example: How Keon Uses Sundays to Run His Content Machine

Keon's a barber and content creator in Philly.

He runs an IG, TikTok, and YouTube channel where he talks about entrepreneurship, grooming tips, and mindset.

Every Sunday, here's how Keon moves:

Step 1: Upload His Weekly YouTube Video to Lately.ai

He copies the video URL or drops the transcript into the dashboard.

Lately breaks that 12-min video into:
- 10–20 short-form social captions
- Suggested post times
- Optimized hashtags
- All formatted for LinkedIn, IG, Facebook, Twitter, and more

Step 2: Review + Edit With His Brand Voice

Lately lets Keon tweak the AI-generated captions to keep that Philly energy and slang.

Prompt:

> *Rewrite this in a confident tone with East Coast flavor. Mention "legacy" and make the CTA feel like a challenge.*

Step 3: Schedule the Week All in the App

He selects his platforms, time slots, and preview window.

Boom. His content is loaded for Monday–Saturday before brunch is even over.

Instructions: How to Use Lately.ai to Build a Week of Content

Step 1: Sign up for Lately.ai

Get your free trial or paid account.

Step 2: Upload a piece of long-form content

This could be:
- A blog post
- A YouTube video (via link)
- A podcast transcript
- An existing social media feed (yes, it can remix *your* past posts too)

Step 3: Let the AI work

Lately will analyze the text, break it into dozens of optimized social media posts, and recommend the best ones based on past performance.

Step 4: Tweak the tone

Use the built-in editor to infuse your brand's language, slang, and emotion. Tell it:

"Make this sound more urgent and culturally relevant to Black entrepreneurs."

Step 5: Schedule it inside Lately

No external apps needed.

Choose post dates, times, platforms, and preview what your audience will see before you hit publish.

Bonus Flow: Make 1 Piece of Content Work 5x

Start with:
- A YouTube video
- A podcast episode
- A written blog or case study

- A new product launch page

Let Lately turn it into:
- IG captions
- Twitter threads
- LinkedIn posts
- Facebook content for your private group

Prompt:

> *Rewrite this video clip into 3 short IG captions. Keep it bold, motivational, and relevant to young Black creatives chasing freedom.*

Sunday Setup Blueprint: 1 Hour to Social Peace

Here's your exact weekly rhythm:

0 AM	w top AI-generated captio

Now your content for the whole week is posted, optimized, and working *while you live.*

The Hustla's Rule:

"Create once. Distribute everywhere. Show up without stressing out."

Lately.ai turns one hour into seven days of brand-building presence.

Because in this season, consistency is louder than hype and automation is louder than burnout.

Talk Data to Me: Analyzing Social Media Performance with AI

Let's be honest… the algorithm don't love you just because you posted.

You can drop the cleanest graphic, the most fire caption, and still feel like it fell into a black hole. That's because **social media isn't just creative, it's calculated.**

You gotta know what time to post.

What content hits hardest.

What your people actually care about and what they scroll right past.

But here's the part they don't teach you on IG Lives and Canva tutorials:

Growth comes from analysis.

And with AI, you don't have to be a data scientist to see the signals.

Step 1: Pull Your Numbers Without the Guesswork

Whether you're using:
- Instagram Insights
- TikTok Analytics
- Facebook Meta Business Suite
- Twitter/X Analytics
- YouTube Studio

… you can copy the data into a spreadsheet or export it, then feed it straight into AI.

Prompt:

> *Analyze this 14-day Instagram Insights report. Identify my top performing content types, average engagement rate, best post time, and areas where I'm falling off. Give it to me in plain English with 3 things I should double down on this month.*

AI Might Say:

- "Carousels had the highest save rate. Keep doing more tip-based content."
- "Your reels are solid but need better hooks in the first 3 seconds."
- "Tuesdays and Thursdays between 11am–1pm had the best reach."
- "Engagement is high, but call-to-actions are weak. Try ending posts with a question."

That's game you can act on.

Step 2: Let AI Build You a Performance Recap

Instead of trying to screenshot your way into clarity, let AI write you a *real* weekly or monthly recap.

Prompt:

> *Create a monthly social media performance report based on this data [paste metrics]. Include reach, engagement, saves, shares, top 3 posts, worst performing content, and suggested action steps for next month.*

Bonus: Use this to present to clients, show growth to investors, or just keep yourself locked in.

Step 3: Compare and Predict

Want to know if your next post idea will hit? Let AI compare it to past data.

Prompt:

> *Compare this new content idea (describe it) to my top 3 performing posts from last month. Will it likely perform well? What should I adjust in tone, format, or caption to maximize engagement?*

Now you're not just posting with hope, you're posting with foresight.

Real-World Example: Monet from Dallas

Monet runs a plus-size lifestyle brand.

Her audience is engaged but picky, what pops one week, flops the next. Before AI, she was overwhelmed by tracking likes and reach manually.

Now?
- She dumps her Instagram Insights into ChatGPT every Sunday
- AI tells her what content made people save, share, and comment
- It even writes her next week's plan based on what hit hardest

Result?

Monet increased her engagement rate by 43% in 2 months

Doubled her email sign-ups by repurposing her best performing content into lead magnets

Started closing brand deals by showing real, data-driven results

The Hustla's Rule:

"Numbers don't lie. They just need a translator."

AI takes the confusion out the metrics and gives you back control. Because when you understand your audience, you stop guessing and start growing with intention.

Talk Data to Me: Analyzing Social Media Performance with AI

Let's be honest, most of us didn't grow up looking at dashboards or spreadsheets.

We were raised on **instinct**.

We post what we feel. We promote what we believe in. We build from the gut.

And while that's powerful, it ain't always profitable.

Because social media success ain't just about creativity. It's about **consistency backed by clarity.**

So if you've ever asked:

"Why did this post do numbers, but the next one flopped?"

"When's the best time to post?"

"Am I even reaching the right people?"

Then this section is for **you**.
And good news? You don't need to be a data nerd.

You just need the *right questions,* and AI will break the numbers down for you like a homie with a whiteboard.

What Are We Even Looking For?

Here's what actually matters when you look at your stats:

- **Reach** – How many people saw your post
- **Engagement** – How many liked, saved, shared, or commented
- **Saves** – Big sign your content was valuable
- **Shares** – Major signal your content was relatable or powerful
- **Click-Through Rate (CTR)** – If you dropped a link, how many people actually clicked
- **Best Posting Time** – When your audience is most active
- **Top Performing Content** – The type of post that hits hardest (reels, quotes, tips, behind-the-scenes, etc.)

You ain't gotta check all this manually.

Just export your Instagram or Facebook analytics once a week or screenshot your

Insights, and let AI break it down for you.

Prompt: AI, Make This Make Sense

Prompt Example:

> *You are a social media strategist for small Black-owned businesses. Here's my Instagram Insights data for the last 7 days [paste or describe]. Tell me which posts performed best, what time I should post, and what kind of content I should stop doing. Keep it simple. I'm not a numbers person.*

AI Might Say:
- "Your carousel post on Wednesday got the most saves. Do more tip-based content."
- "Your quote graphic reached 1,200 people but had low engagement. Add a stronger caption next time."
- "Reels got 4x more shares than photos. Lean into motion."
- "Best post times were 11 AM and 8 PM. Try to stay consistent with that window."

Boom. Now you've got direction, not confusion.

How to Get Your Data (Simple Version)

On Instagram:
1. Go to your profile
2. Tap **"Professional Dashboard"**
3. Hit **"See All Insights"**
4. Tap "Last 7 Days" or "Last 30 Days"
5. Screenshot the overview OR copy the numbers manually into a Google Sheet

On Facebook:
1. Go to your Facebook Page
2. Click **Meta Business Suite**
3. Select **Insights** from the left menu
4. Export your Page Summary or just jot down the top 3 posts and times

Prompt: Build Me a Weekly Performance Report

Prompt:

> *Based on this week's Facebook and Instagram data [paste numbers], give me a simple weekly report: top post, worst post, best time to post, overall engagement rate, and 3 action steps I should take next week to improve.*

AI Might Reply:
- "Your Monday Reel got the most engagement. Post similar content every Monday."
- "Quotes got more shares than tips. Lean into emotional content."
- "Your 8 PM post had double the CTR than your 12 PM one. Shift your schedule."
- "Try a call-to-action that invites DMs instead of just link clicks."

That's how you **adjust with confidence.**

Real-Life Example: Kayla the Candle Boss

Kayla runs a handmade soy candle business out of Baltimore. Before AI, she was posting whenever she felt like it. Sometimes at 2 AM. Sometimes during lunch. Some posts would do great. Others? Crickets. She had no clue why.

Now?

Every Sunday she screenshots her IG Insights, pastes them into ChatGPT, and says:

"Tell me what worked and what didn't. Help me plan this week."

AI tells her:
- Which product videos performed
- What content to double down on
- What caption tone pulled the most DMs
- And what time her audience is really locked in

Kayla went from 200 views per story to 2,000

From inconsistent content to a weekly plan that works

From "Why didn't this work?" to "Let's run that play again"

The Hustla's Rule:

"Data ain't just for tech bros and marketers, it's your digital truth."

Let AI decode the numbers so you can stop guessing and start growing.
Because when you know what works, **you don't hustle harder, you hustle smarter.**

Grow the Fam: Using AI to Boost Followers and Spark Real Engagement

Followers don't pay the bills, but the right community will fund your freedom. It's not just about having a big following; it's about building the right *audience*. People who vibe with your mission. Rock with your brand. Click, share, comment, and **convert**.

And real talk?

You don't need to chase clout. You need to build *connection*. Because when the connection is real, the algorithm gotta respect it.

This section shows you how to use AI to grow the fam organically, strategically, and with content that sparks conversation, not just attention.

Step 1: Attract the Right People with the Right Energy

You don't need everybody; you need *your people*.
AI can help you find them and speak directly to their head *and* their heart.

Prompt:

You are a culturally fluent copywriter for Black-owned wellness brands. Write 3 Instagram bio options for a yoga teacher who helps Black men reconnect with their bodies and their peace. Keep it powerful, concise, and magnetic.

AI Delivers:
- "Peace is the flex. Certified yoga coach for kings reclaiming their power."
- "Helping Black men move through trauma, one breath at a time."
- "Strength, stillness, and softness on and off the mat."

Your *bio* is your billboard. Let AI sharpen it up, so folks know *instantly* what you stand for.

Step 2: Use AI to Write Comment-Ready Captions
Captions shouldn't be cute; they should be **conversation starters**.

Prompt:

Write 5 Instagram captions that ask engaging questions for a financial coach helping single moms build generational wealth. Tone: motivating, non-judgmental, and culturally relevant.

AI Might Say:
- "If your money could talk, what would it say about your boundaries?"
- "What's the first lesson about money you want to teach your kids?"
- "Drop a   if you're the first in your family breaking the paycheck-to-paycheck cycle."

This ain't content, it's *community building*.

Step 3: Spark DMs & Comment Sections with Callbacks

Prompt:

Write a Reel caption that ends with a question designed to boost comments. Audience: Black Gen Z creatives navigating burnout. Topic: protecting your peace while pursuing your purpose.

Example:

"The grind can't cost you your glow. What's your go-to move when you feel burnout creeping in?   Let's talk about it."

Why this works:

You're inviting stories. Sharing space. Turning a post into a *conversation*.

Step 4: Turn Engagement Into Momentum

Let AI help you build on your wins.

Prompt:

> *Based on these top-performing posts [insert topics or headlines], create 5 new post ideas with similar themes that invite shares, saves, or comments.*

AI might respond with:
- "Turn your popular carousel into a short-form video"
- "Create a 'myth vs fact' version of your last post"
- "Host a live Q&A around your highest-comment topic"
- "Turn your tweet-style post into a 3-part series"

AI keeps you in your lane, but levels up your *strategy*.

Bonus Prompt: Growth Strategy Plan

Prompt:

> *Build a 30-day engagement growth strategy for a Black woman-owned candle brand. Focus on growing Instagram followers and boosting DMs. Include post types, engagement CTA ideas, and stories that spark interaction.*

You'll get:
- Daily post types
- DM strategies
- Comment tactics
- Weekly engagement themes

- CTA templates

And now your growth plan ain't a mystery, it's a *mission*.

The Hustla's Rule:

"You don't need a blue check to build a real tribe."

Use AI to attract, connect, and convert with content that don't just look good, but *feels good*. Because when your message speaks to the soul, the numbers follow the vision.

Crisis Ain't Gotta Be Chaos: AI Tools for Dealing with Trolls, Drama & Digital Smoke

If you're building something real, the hate gonna come with it. Not "if," but **when.**

And the bigger you get, the louder the noise:

Folks mad at your prices
Customers wildin' in your comments
Trolls trying to go viral off your name
People stealing your content or calling your whole brand into question

And for Black entrepreneurs? It hit different.

We're building in public.

We're learning in real time.

And we don't always have PR teams or crisis managers ready to clean it up.

That's why this section is *critical.*

This is about **emotional intelligence meets AI defense systems.**

First Rule: Protect the Brand, Protect Your Peace

Before you respond with feelings, respond with *focus.*

Let AI handle the tone. You just bring the truth.

You don't need to stoop. You need to **stand firm**.

AI helps you take the emotion *out* the first draft, so your response hits like a statement, not a breakdown.

Hater-Blocking Blueprint: How to Handle Digital Heat with AI

1. The Classy Clapback (aka Boundary-Setting)

Prompt:

> *Write a calm, confident response to someone accusing me of being "too expensive" on my small business page. I want to address it without feeding the drama and still stand on my worth. Tone: Graceful but firm. Audience: Black professionals and culture-conscious customers.*

AI Might Say:

"Our prices reflect the time, quality, and value we put into every product. We honor those who recognize that, and we invite those who don't to support businesses that fit their expectations. Either way, peace."

That's how you silence hate with elegance.

2. When a Mistake Happens and You Gotta Own It

Prompt:

> *Draft a short IG and Facebook apology post for a shipping delay that affected 75 customers. I want to be accountable, keep it real, and remind people we're still a growing business committed to excellence. Make it warm, honest, and strong.*

AI Might Say:

"We missed the mark this week, and we want to say thank you for your patience. As a small but growing brand, every order matters. We're already fixing it behind the scenes, and your support means the world as we keep leveling up."

Bonus Tip: Include a link to a FAQ or resolution page to reduce DMs and confusion.

3. Handling Negative Reviews Without Getting Defensive

Prompt:

Write a professional, culturally sensitive reply to a negative review where the customer claims they had a bad experience, but I believe the issue wasn't on my end. Help me stand on my business without arguing.

AI Might Say:

"Thank you for your feedback. We hear you. Our goal is always 5-star service, and we're sorry to hear this didn't meet your expectations. If you're open, we'd love the chance to talk and make it right. DM us or email support@yourbrand.com. Let's grow together."

That shows the **brand is solid even when it's tested.**

4. Build a Crisis Toolkit with AI: *Before* You Need It

Your brand needs a go-bag for drama. Create a ready-made folder with:

- Pre-written replies for common situations
- A "media response" statement for public call-outs
- A refund & return policy in plain English
- A DM script for when things get spicy privately
- A pinned post ready to drop when people start asking questions

Prompt:

Build me a social media crisis toolkit for a Black woman-owned wellness brand. Include templates for: shipping delays, pricing complaints, influencer drama, and customer service mix-ups. Tone: grounded, protective, and rooted in integrity.

Now you got options. And you're never caught slippin'.

Real-Life Example: Bri the Baker

Bri runs a bakery in Atlanta. One weekend, she got dragged on Twitter because she turned away a last-minute customer after closing. Someone recorded it, cut out context, and posted it like she was rude. The post went viral. People were in her comments talking wild.

Bri didn't react emotionally. She:

1. Prompted AI to help her write a community statement explaining her hours + boundaries
2. Posted a Reel the next day showing her team cleaning up and why they close early
3. Sent a follow-up email to her customer list reinforcing her values and thanking folks for still showing love

She flipped the narrative.

Gained 5,000 new followers

Got featured in a "Support Black Bakers" IG roundup

Didn't fold when the pressure came

Bonus Hustla Prompts

Write a "we hear you" statement for when your IG comment section gets flooded with complaints.

Draft a brand boundary post for when people ask for discounts or freebies. Keep it respectful, but empowering.

Create a mini FAQ that I can pin to stories when folks are upset or confused about policy changes.

The Hustla's Rule:

"You don't owe every critic your energy, but you do owe your brand a response that's rooted in strength."

Let AI help you lead with grace, move with clarity, and keep your name clean, even when the net tries to dirty it up.

Because real bosses don't panic; they position.

Chapter Eight

Brandz N' The Hood
Managing Your Brand in the Age of Algorithms

Listen, anybody can make a logo.

Anybody can run a page, buy some followers, and fake it for a few weeks. But a *brand*? A real brand? That's deeper. A brand is how your people talk about you when you're not in the room. It's the energy that lingers after the sale. It's the message *behind* the merch.

And most of all; it's **consistency**.

In the hood, we build reputation one move at a time.

In business? Same rules apply.

This chapter is about using AI to protect that reputation, **tighten your voice**, keep your message consistent across every platform, and help your brand grow with intention and not confusion.

What We're Gonna Cover in This Chapter:

- How to train AI to speak in your brand voice (so your content stays on code)
- Tools to keep your branding consistent from your captions to your contracts
- Building out brand guidelines using AI (even if you never hired a designer)
- Monitoring your brand reputation and responding with poise
- Evolving your message without losing your identity

Because when you're coming from where we come from, **your name is your currency**.

And out here? Your brand *is* your name.

The Hustla's Reminder:

"If you don't define your brand, the internet will do it for you."
Let AI help you keep it clear, keep it clean, and keep it connected to the mission. This ain't branding for clicks, it's branding for **legacy**.

Talk Like You Walk
Defining and Refining Your Brand Voice with AI

Your logo might catch attention, but your voice?

That's what builds loyalty.

Your brand voice is how you show up in the world; online and offline. It's the energy behind every caption, every email, every product description, every bio. It's the *tone* that makes people say, "Oh yeah, that's definitely them."

But here's the problem:

Most entrepreneurs write different every time they post.

One day it's professional. The next day it's all slang. The next day it's just… confused.

And confusion don't convert.

That's why this section is all about using AI to **lock in your brand voice and stay on message.**

Step 1: Discover Your Voice (If You Don't Know It Yet)

If you've never defined your tone, start with these 3 questions:
1. How do you want people to *feel* when they read your content?
2. If your brand was a person, how would they talk?
3. What 3 brands (or artists, influencers, movements) share a similar vibe?

Use this info to **train AI to write like you.**

Prompt:

> *You are a voice branding expert. Help me define my brand voice. My business is a wellness brand for Black women. I want to sound soulful, grounded, empowering, and culturally fluent. My inspirations are Solange, Erykah Badu, and Hey Fran Hey.*

AI Might Return:
- Tone: Calm, lyrical, affirming
- Language: Conversational, rooted in Black culture, rich with imagery

- Pacing: Intentional and rhythmic
- Emotional triggers: Healing, identity, self-love, ancestral connection

Now that's a blueprint.

Step 2: Train AI to Write Like You

You can literally *feed AI your tone* and make it write on your level.

Prompt:

> *Here are 3 past captions I wrote [paste them]. Based on these, define my brand voice and use it to write a caption for my next product drop. Audience: Black millennial women who want luxury without the guilt. Tone: real, soft, and a little bougie.*

AI learns from you.

Then it replicates with **your sauce**, not some generic AI energy.

Step 3: Build a Brand Voice Cheat Sheet (with AI)

If you ever hire a VA, bring on a social media manager, or just want to stay consistent, create a brand voice doc.

Prompt:

> *Build a 1-page brand voice guide for my candle company. We blend hip-hop culture with luxury aromatherapy. Include: tone, slang we use, phrases to avoid, emojis we love, and sample CTAs.*

Now your whole team is on the same page; whether it's you writing or the machine.

Real-Life Example: Malik the Media Coach

Malik helps creatives land brand deals. His voice is 50% Nipsey Hussle, 50% Gary Vee. Before using AI, he felt like every caption sounded different.

Now?
- He gave ChatGPT old IG posts
- Trained it to understand his motivational but gritty tone

- Has AI write captions, emails, and even ad copy that sounds just like him

Result? Engagement up. Consistency locked in.

Malik got more time to coach and less time battling blank screens.

Bonus Prompt: Voice Switcher

Rewrite this email in 3 different versions: 1) corporate formal, 2) urban professional, and 3) motivational mentor. Keep the message the same but switch the tone for each.

Why? Because sometimes your voice needs to shift depending on who you're talking to.

AI helps you pivot without losing your identity.

The Hustla's Rule:

"When your voice stays consistent, your brand becomes unforgettable."

AI can help you speak clearly, confidently, and in alignment with your mission every time, every post, every piece of content. Because the strongest brands don't just talk, they talk with *presence*.

Stay Solid: Keeping Your Brand Consistent Everywhere

When people land on your IG, your website, your email, or your packaging, **they should feel like they're stepping into your world.**

That's what consistency does.

It builds trust.

It makes you look organized, polished, and legit, even if you're running your business from a laptop on the kitchen table. But when your tone is one way on Monday, your visuals look different on Wednesday, and your email signature sound like a stranger wrote it?

That's brand confusion. And confusion don't convert.

This section shows you how to use AI to keep your brand **tight, recognizable, and real**, from colors and fonts to captions and catchphrases.

Step 1: Create a Brand Consistency Guide with AI

If you've never made a style guide or brand kit, no worries, AI can help you build one in minutes.

Prompt:

> *You are a brand strategist. Create a brand style guide for a streetwear brand inspired by 90s hip-hop and Black radical art. Include tone of voice, color palette inspiration, typography vibe, logo usage notes, and brand keywords. Keep it tight and street but creative.*

AI Might Return:
- Tone: Bold, unapologetic, nostalgic
- Color palette: Golds, blacks, earth tones
- Fonts: Retro block fonts mixed with handwritten script
- Keywords: Culture, resistance, fashion, legacy
- Tagline suggestions: "Built for the Culture. Worn by the Movement."

Now you've got a blueprint to stay on-brand *everywhere*.

Step 2: Use AI to Check Your Consistency

You can even use AI like a brand auditor.

Prompt:

> *Analyze these 3 social posts, this email, and this product page [paste or describe]. Tell me if my voice and visuals feel aligned. What should I adjust to make it feel more consistent across the board?*

Let AI catch the vibe breaks you might miss.

Step 3: Keep Visuals Tight with AI + Canva

If you're designing flyers, ads, or social posts, AI can help you describe what you want, and **tools like Canva can help you lock in templates,** so everything matches.

Prompt:

Describe a Canva brand kit for a feminine but high-energy coaching brand run by a Black woman. I want my visuals to look confident and polished without feeling stiff. Suggest colors, font pairings, and layout ideas for carousels and IG Stories.

You'll walk away with a whole design vibe without ever opening Photoshop.

Pro Tip: Once you've got your brand kit locked in, duplicate your best templates for:

- Stories
- Quote graphics
- Product drops
- Live reminders
- Testimonials

Now every time you post, it looks like you hired a whole marketing team. (But it's just you... and your AI.)

Real-Life Example: Chanel the Coach

Chanel's a life coach for Black women navigating corporate burnout.

Her content used to feel all over the place, sometimes pink and bubbly, sometimes all beige and quotes.

Once she built a brand kit with AI + Canva, everything aligned:
- Color scheme = soft rose, gold, and charcoal
- Fonts = modern serif with a touch of script
- Voice = soft-spoken, affirming, and no fluff
- IG, email, and podcast visuals = *uniform and unmistakably hers*

Result? Her brand now **feels like a safe space**, no matter where you meet it.

Bonus Prompt: Brand Consistency Checklist Builder

Build a weekly checklist for brand consistency across social, email, and website. Include what to review visually and verbally before publishing any new content. Make it simple enough for a solo entrepreneur to use every Sunday.

Now your Sunday reset includes:
- Voice check
- Font and color check
- CTA alignment
- "Does this feel like me?" gut-check

The Hustla's Rule:

"Your brand ain't just what you say, it's how you say it, show it, and repeat it."

Let AI help you lock in your identity, tighten your visuals, and stay on-brand even when you're busy building.

Because real brands aren't random, they're **rooted.**

Eyes on the Block: Monitoring Your Brand Reputation with AI

Let's keep it 100:

In this era, your brand could be trending and tarnished in the same 24 hours. And if you ain't paying attention? You could lose money, momentum, and your message before you even realize something popped off.

That's why monitoring your brand reputation is *non-negotiable.* You built this name brick by brick; **you can't afford to let it get played with.**

But here's the good news:

You don't need a whole PR team. You've got AI.

And AI don't sleep.

Step 1: Use AI Tools to Watch the Conversation

There are platforms that monitor when your brand name is mentioned across social media, blogs, YouTube, and even reviews.

Top Free or Low-Cost Tools:
- **Google Alerts** – Get emails anytime your name or business is mentioned online

- **Brand24** – Tracks mentions, sentiment, and engagement across platforms
- **Talkwalker** – Gives insight into where and how you're being talked about

Prompt (for ChatGPT + Data Tools):

Analyze the past 7 days of mentions for my brand "PowerGlow Skincare." Tell me if the sentiment is mostly positive, neutral, or negative. Pull key words used in the mentions and recommend if I need to respond or let it ride.

AI Might Say:
- "Positive sentiment around your latest serum launch"
- "One customer complaint gaining traction, recommend addressing with a comment and pinned story"
- "Your name was tagged in a trending thread about Black-owned beauty brands. Capitalize on that attention with a repost + thank you post"

Now you're not reacting emotionally. You're **responding strategically.**

Step 2: Let AI Help You Respond with Class

Prompt:

Write a calm, on-brand response to someone who gave us a 2-star review for slow shipping. I want to show empathy, explain we're improving, and invite them back with a discount code.

Example:

"We're grateful you gave us a shot, and we hear your frustration. We're actively improving our shipping system and appreciate your patience as a growing brand. DM us for a code. We'd love to earn your trust back."

AI helps you sound like a CEO and not a comment section troll.

Step 3: Identify Brand Misuse or Imitation

Prompt:

Search online mentions of "Culture Kicks Apparel" and alert me if another company is using a similar name, selling similar designs, or tagging my products without credit. Suggest how I should handle it legally and publicly.

Pro Tip:

Follow that up with:

"Draft a cease-and-desist email template and a public response that's professional but protective of our brand."

Because sometimes it's not shade, it's theft. And your IP deserves to be guarded.

Step 4: Track What Content Drives Brand Buzz

Prompt:

Analyze my last 30 posts. Tell me which ones got the most shares, tags, and outside mentions. Break down what made them buzz-worthy and how I can remix them for future content.

This ain't just about damage control, it's about growth control. Find what *sparks conversation* and double down on it.

Real-Life Example: Jada the Juice Queen

Jada runs a Black-owned cold-pressed juice business outta Detroit. One day, a fitness influencer posted about her with no tag, and people started ordering from a knockoff using a similar name.

Instead of panicking, she:
1. Set up a Google Alert with her brand + city
2. Used ChatGPT to draft a respectful but protective message to the influencer
3. Posted a story clarifying her brand and inviting people to the real page
4. Asked AI to help her plan a campaign using the attention to her advantage

Result: She got the tag, tripled her followers that week, and dropped a "Juice Queen

Verified" campaign that turned clout into conversions.

Bonus Prompt: Brand Monitoring Setup Checklist

Build me a weekly brand monitoring checklist. Include tools I should check, keywords I should track, and questions I should ask AI each Sunday to stay ahead of drama and opportunities.

Checklist Might Include:
- Check Google Alerts for new mentions
- Run a sentiment check on mentions and comments
- Review social tags and DM folders
- Use AI to write responses or highlight what needs public attention
- Ask: "What's the public vibe around my brand this week?"

The Hustla's Rule:

*"Don't just build a brand, **guard it.**"*

Let AI be your digital watchdog. Because your name carries weight, and you worked too hard to let a comment, clone, or misstep play with your legacy.

Stay aware. Stay protected. Stay ready.

Evolve Without Losing the Essence: Letting Your Brand Grow Right

Every real brand gotta grow.

What you started with may not be what you stay with, and that's okay.

The message might expand.

The visuals might get sharper.

The audience might shift.

But the mission? That stays rooted.

Too many entrepreneurs stay stuck in Version 1.0 because they scared folks won't follow the update. And on the flip side? Some pivot so hard that their brand don't even feel like them anymore.

This section is about how to use AI to help you evolve *gracefully*.

To scale up without selling out.

To glow up without going ghost on your identity.

Step 1: Ask AI to Audit Your Current Brand Identity
Before you evolve, you gotta know where you stand.

Prompt:

> *Act as a brand consultant. Based on these IG captions, my About page, and my last email blast [paste or summarize], tell me how my brand voice, visuals, and message currently come across. Is it aligned with who I'm becoming?*

AI Might Say:
- "Your content feels grounded but leans heavy on hustle. As you evolve into thought leadership, consider softening the tone with more storytelling."
- "Visuals are strong but need consistency. Fonts and filters vary too much across platforms."
- "You're attracting early entrepreneurs, but your offer is better suited for growth-stage businesses."

That insight is gold.

Step 2: Write a Brand Evolution Statement

If you're making a shift, let your people know. Don't go ghost and pop up rebranded like Beyoncé in a hoodie.

Use AI to help you craft the rollout.

Prompt:

> *Write a 2-paragraph brand evolution post for Instagram. I'm shifting from done-for-you marketing services to digital education for Black business owners. Keep the tone empowering, rooted, and honest. Help me speak to my day-ones while inviting new folks to the table.*

Example:

"This ain't the end, it's the expansion. What started as a service has grown into a mission: to equip more of us with the tools to build and scale with confidence. We're stepping into a new chapter, and I'm grateful to bring y'all with me."

Step 3: Let AI Help You Rebrand (Without Starting Over)

You don't need to burn it down to build something better. AI can help you remix what's working and align it with where you're headed.

Prompt:

> *Help me rewrite my mission statement and tagline. I want to reflect a new focus on social justice through fashion, without losing the streetwear edge. Keep it short, bold, and purposeful.*

AI Might Return:
- Mission: "We design with purpose. Wearable protest rooted in culture and truth."
- Tagline: "Not just fashion. A statement."

Now your *next* feels connected to your *now*.

Real-Life Example: Maya the Movement Builder

Maya started as a personal trainer. Her brand? "Built to Flex." Over time, her work evolved into mental health, nutrition, and spiritual wellness for Black women. She knew "Built to Flex" was only telling part of the story.

Here's what she did:
1. Had AI audit her content and point out which messages people engaged with the most
2. Used that info to shift her brand voice from high-energy fitness to holistic wellness
3. Rewrote her tagline: from "Train Hard. Live Loud." to "Strong. Soft. Sacred."
4. Created a 7-post IG rollout (with AI's help) explaining the shift and launching her first journal

She retained 90% of her old audience

Gained 4,000 new followers

Sold out her first drop in 72 hours

She didn't lose her essence, she **leveled up**.

Bonus Prompt: Brand Glow-Up Blueprint

Help me plan a full brand evolution rollout across IG, email, and website. I'm shifting my focus from solo freelancers to creative agencies. Include: a public post, an email to my list, a new mission statement, and tagline ideas.

Now you're not just changing, you're curating a whole experience.

The Hustla's Rule:

"Growth is the goal, but alignment is the requirement."

AI helps you evolve your message, expand your impact, and stay rooted in the reason you started.

So when you glow up, your people recognize you; even in your next-level form.

Chapter Nine

Picture Me Rollin'
AI Powered Visuals That Sell the Vision

Let's be honest, if it don't *look* good, most folks won't stop to see if it's good. You could have the dopest service, the deepest mission, the best product in the city...

But if your brand visuals look sloppy, last-minute, or inconsistent?

You lose trust.
You lose clicks.
You lose the bag *before they even hear your message.*

In a world that scrolls fast and forgets faster, **how you show up visually matters.** And guess what? You don't need to be a graphic designer to build visuals that stop thumbs and start conversations.

You just need AI.

And a system.

And this chapter.

What We're About to Cover:
- How to use AI tools to create logos, flyers, social media graphics, and branded templates
- Turning rough ideas into polished visuals using Canva + AI
- Generating branded imagery from scratch, even if you don't have product photos yet
- Building a reusable visual identity system (so everything you post *feels like you*)
- Using visuals to tell your story, elevate your offer, and look like money *before* the money comes

Because first impressions ain't just visual, they're *visceral.*

People buy with their eyes first.

If it looks sloppy, they think the product is sloppy.

If it looks intentional? They believe the brand is solid.

The Hustla's Reminder:

"Your visuals should reflect your value."

AI gives you the power to design like a pro, even when your resources say, "we still bootstrappin'."

This ain't Canva hustle. This is visual storytelling with **vision**. Let's roll.

Design Ain't Optional: Creating Branded Visuals with AI

Let's make one thing clear:

In today's game, **design is marketing.** If your post don't pop in the first second, they're already gone. If your flyer looks like a high school project, they're scrolling right past it. And if your visuals don't *feel* like your brand? You're blending in when you should be standing out.

But here's the good news:

You don't need Photoshop.

You don't need a designer on retainer.

You just need the right prompts and the right tools.

This section is your starter kit for using AI to create:
- Logos
- Social media graphics
- Flyers
- Brand kits
- Product mockups
- And visuals that match your message **every time**

Step 1: Turn Your Vision into Mockups Using AI

Sometimes, the idea is in your head, but the execution feels out of reach.

Start with image-generation tools like:
- **DALL·E** (built into ChatGPT)
- **Canva AI (Magic Media)**
- **Midjourney** (for high-end visuals)
- **Davinci** (for AI-enhanced image and video assets)

Prompt:

Generate an Instagram-ready graphic of a Black woman entrepreneur sitting in a modern home office, reviewing client notes. The tone is aspirational, soft luxury. Use warm neutral colors, sunlight, and branded elements (journal, candles, laptop).

Now you've got a visual that fits your vibe without needing a camera or crew.

Step 2: Use AI + Canva to Build Branded Graphics Fast

Once you've got your tone and aesthetic, build out a few **go-to templates** in Canva.

Let AI help with:
- Copywriting
- Color palette suggestions
- Font pairings
- Layouts for stories, carousels, promos, and testimonials

Prompt:

Create 5 Instagram post ideas for a life coach helping Black women reclaim their time. Include suggested headlines, colors, and design formats (carousel, quote, tips, video preview).

AI Might Give You:
- Post 1: "3 Signs You're Doing Too Much" (carousel, bold font, pink/gold palette)
- Post 2: Quote graphic: "Rest is a revolutionary act."
- Post 3: Client story teaser with sage green background
- Post 4: Webinar promo video
- Post 5: Poll post: "Time freedom vs. Financial freedom"

Design it once. Duplicate forever.

Step 3: Use AI Mockups to Sell Before You Even Launch

No professional photos yet? No problem.

Use AI to generate **product mockups** and branded visuals that look ready for the shelves.

Tools to Use:
- **DALL·E** for lifestyle visuals
- **Midjourney** for luxury product renders
- **Canva Smartmockups** for product previews
- **Davinci** for cinematic-style visuals and AI-enhanced product videos

Prompt:

> *Generate a clean mockup of a Black-owned skincare line with minimalist packaging. Label says, "Glow Queen Serum." Place it on a marble bathroom counter with sunlight and greenery.*

Now you've got ad-ready visuals without booking a shoot.

Real-Life Example: Nate the Notary

Nate launched a mobile notary service in Charlotte with zero visuals.

He used AI + Canva to:
- Generate his logo
- Build branded IG templates
- Use DALL·E to create photos of a Black male notary at work
- Use Canva Smartmockups to showcase branded folders and notebooks
- Build a clean one-pager that looked like corporate gold

30 days later: 15 clients, a law firm partnership, and a brand that *looked* like it had been running for years.

Bonus Prompt: Visual Identity Kit Builder

> *Build a visual identity kit for a luxury men's grooming brand targeting professional Black men. Include color palette, font pairings, image direction, and post format suggestions for IG, LinkedIn, and product packaging.*

AI Delivers:

- Colors: Charcoal, brushed gold, mahogany
- Fonts: Serif headline + clean sans-serif body
- Imagery: Dim lighting, upscale grooming, gold accents
- Tagline: "Polished. Powerful. Proven."

The Hustla's Rule:

"If your content don't catch the eye, it won't reach the mind."

Let AI help you create visuals that validate your value, sharpen your identity, and sell the story before they even read the caption.

Because the brand ain't just what you say, it's what you *show*.

Remix the Content: Stretch One Idea into Ten with AI

You ever feel like you need to post 5x a week but don't have 5x the time or creativity?

Here's the truth:

You don't need more content. You need smarter content.

The biggest brands don't create nonstop; they **repurpose like pros.** One video becomes a blog post. One blog becomes an email. One tweet becomes a Reel script. That's not lazy, that's leverage.

And now with AI? You can take one idea and turn it into a whole **visual ecosystem** that works across Instagram, Facebook, LinkedIn, your email list, and your product pages.

Step 1: Feed AI One Core Piece of Content

Start with a blog, a podcast, a YouTube video, a long-form IG caption, or even a customer review.

Prompt:

> *Take this blog post [paste or link] and turn it into a week's worth of content. I want 2 IG captions, 1 quote post, 1 Reel script, 1 carousel idea, and a story question.*

Audience: creative entrepreneurs who follow my brand for inspiration and marketing game.

AI Might Deliver:
- **Caption 1:** Break down one key point with a personal story
- **Caption 2:** A stat + CTA to learn more
- **Quote Post:** "Your content shouldn't work harder than your system."
- **Reel Script:** Quick tip on how to automate your content calendar
- **Carousel:** "5 Ways to Repurpose 1 Piece of Content"
- **Story Question:** "What platform are you sleeping on right now?   "

All from **one idea.** That's content efficiency.

Step 2: Build a Multi-Platform Flow with AI

AI doesn't just write; it can **strategize where each piece belongs.**

Prompt:

Based on this blog post and my brand voice, tell me what parts to post on IG, which ones to use for email, and what should be turned into a LinkedIn post. Then help me write each one.

Bonus: Use tools like **Canva, Davinci,** or **DALL·E** to generate visuals for each piece. AI handles the words, visuals, and layout. You just upload and go.

Real-Life Example: Alicia the Artist Coach

Alicia runs a coaching program for Black visual artists. She used to write one fire caption per week and let it stop there.

Now she:
1. Takes that caption and feeds it to AI
2. Gets 2 quote cards, a story question, a tip carousel, and a short email version
3. Loads everything into Canva and schedules it with Later
4. Spends Sundays sipping wine while her week's content is already loaded

Result: 3x engagement, new clients, and no burnout

Bonus Prompt: 10x Repurposing Engine

Take this YouTube video [insert link] and turn it into a content plan. I want:
- *3 Instagram captions*
- *1 short-form video script*
- *1 Twitter thread*
- *1 email blast*
- *1 LinkedIn post*
- *1 Pinterest graphic idea*
- *1 blog intro*
- *1 call-to-action for my coaching program*

Now you're everywhere **without *being* everywhere**

The Hustla's Rule:

"You don't need more content; you need better circulation."

Let AI turn one idea into 10 moves. That's how you show up more, do less, and stay consistent.

This ain't content creation, it's content compounding. Let it grow. Let it go viral. Let it work.

System Over Stress: Building a Visual Content System with AI

Posting every now and then when inspiration hits? That's cute. But if your brand is the business, **your visuals need to move like a machine.**

The difference between a hobby and a hustle is **a system.**

In this section, we're building a **visual content system** that's:
- Branded
- Automated
- Scalable
- And *so clear* you could pass it off to a VA or let AI run it solo

Let's get you out of "content chaos" and into **content clarity.**

Step 1: Create a Visual Content Template Vault

Start by building your go-to templates in Canva for:
- IG Stories
- Reels Covers
- Quote Graphics
- Product Highlights
- Testimonials
- "Coming Soon" or Drop Announcements

Prompt:

Create a visual content checklist for a fashion brand launching new collections monthly. Include template types, captions, CTA suggestions, and colors to match a bold, Black streetwear vibe.

AI Might Deliver:
- Carousel template for storytelling
- "Tap in now" CTA-style reels cover
- Quote card template: black & gold w/ graffiti font
- "Drop Date" template: countdown format
- Testimonial card w/ street-style photo layout

Now every piece you drop feels intentional, not random.

Step 2: Set a Weekly Visual Rhythm (Sunday or Not)

Every Sunday, or whatever day you breathe, you run this system:

Task	**Tool**	
Auto-schedule posts	Lately, Canva, or IG Scheduler.	

In under an hour, you've built a *week of visual marketing.*

Step 3: Track What Works and Let AI Adjust the System

Let AI help you optimize your system over time.

Prompt:

Here are my last 20 posts and their stats [summarize or paste]. Tell me which visuals drove the most engagement, what style or post type I should replicate, and where I can improve my CTA game.

AI might say:
- Your bold-colored quotes had the highest shares
- Posts with testimonials got saved more than tip carousels
- You need stronger end slides on carousels. Add clearer CTAs

Now your system **evolves itself** as you grow.

Real-Life Example: Dee the Doula

Dee runs a doula and wellness brand for Black moms. She was posting randomly, writing from the heart, but it wasn't consistent, and her visuals didn't match her vibe.

She used AI to:
- Build 6 reusable Canva templates
- Create a 4-week visual content calendar based on her services
- Generate consistent captions, colors, and story themes
- Use Davinci to produce a 15-second animated Reel intro with her brand name

Now she moves with **peace and presence.**

More inquiries. More shares. Less stress.

Bonus Prompt: Visual Content System Builder

Build me a plug-and-play visual content system for a Black-owned candle brand. I post 4x per week. I want templates, post types, suggested AI tools, and a weekly batching flow. Keep it realistic for a one-person team.

AI Might Say:
- Monday: Quote Graphic (Canva)
- Wednesday: Product Demo (Davinci-generated video)
- Friday: Customer Testimonial (Smartmockup + Canva)
- Sunday: "Self-Care Tip" Carousel (AI-written, Canva-designed)

The Hustla's Rule:

"Freestyle is for mixtapes, not for marketing."

Systems let you post with purpose, show up consistently, and look like a brand even when you're building solo.

Let AI be your co-creator and Canva your canvas. The visuals? They gon' speak for themselves.

Visuals That Close: Turning Your Content into Cash

Let's keep it a buck:

Pretty don't pay the bills, but **strategy does.**

You could have the dopest flyer, the cleanest IG grid, the crispiest mockup on the timeline... But if you ain't telling people what to do next? You just got decoration, not marketing.

This section is about turning visuals into **calls to action**, into **movement**, into **money**.

Because your posts shouldn't just get likes, they should *get booked, get clicked, get sold out.*

Let's get your visuals to **do the heavy lifting**.

Step 1: Make Every Visual Work the Funnel

Every piece of content should do at least one of three things:
1. **Attract** (new people into your world)
2. **Nurture** (build trust with folks already following)
3. **Convert** (turn viewers into buyers or leads)

Prompt:

> *Rewrite this Canva carousel into three versions: one that attracts new followers, one that nurtures my audience, and one that pushes a product sale. My brand is a handmade Black-owned jewelry line for spiritual women.*

AI Might Return:

- Attract: "5 Crystals That Help You Protect Your Energy (Swipe)"
- Nurture: "Here's why I started hand-making every piece myself…"
- Convert: "This new 'Sacred Armor' piece just dropped. Limited stock. Tap in."

Same vibe. Three angles. Three levels of value.

Step 2: Drop CTAs That Actually Move People

CTAs = **Call to Action**. And no, "link in bio" ain't always enough.

Prompt:

> Give me 10 CTA lines I can use at the end of visual posts that sound more conversational and culturally aligned. Audience: Black women entrepreneurs.

Examples:
- "Ready to invest in yourself? Let's build."
- "Don't just scroll, tap into what's yours."
- "This feel like you? DM me 'info' and I'll send the details."
- "If this hit your spirit, drop a   and share it with the tribe."
- "Your soft life starts with smart moves. Let's go."

These don't just sell, they **speak.**

Step 3: Use AI to Design Sales-Focused Visuals

Need a promo slide? Launch flyer? "Last chance" post? AI's got you.

Prompt:

> Design a 3-post visual sales campaign for my coaching program. Include a countdown post, a testimonial post, and a value post that breaks down why my method works. Use a confident tone and keep the visual flow branded.

Now plug that into Canva or Midjourney, add your logo, and you've got a mini sales funnel running through your grid.

Step 4: Move People Off the Feed and Into the Bag

Social media is the door. Your offer is the house.

Prompt:

> *Write a caption for my Reel that invites people to join my waitlist without sounding salesy. Mention transformation, limited spots, and link in bio.*

✅ Example:

"This work ain't just about scaling, it's about reclaiming your time, your value, and your peace. Doors opening soon. Limited seats. If you know you're ready, you already late. Link in bio."

Visuals bring them in.

Words close the deal.

👤 **Real-Life Example: Tiff the Tech Educator**

Tiff teaches tech skills to Black high school girls. She wanted to fill her free summer coding bootcamp but didn't want to sound too salesy or corny.

She used AI to:
- Create a 3-part IG visual series breaking down why tech = freedom
- Design flyers in Canva + Midjourney with her colors and message
- Use AI to write captions that called out her community with pride, not pressure
- Set up a CTA that said: "Tag a young queen who needs to see this 👑"

💥 40 sign-ups in 4 days
💥 2 local news outlets reached out
💥 She turned visuals into **visibility and value**

Bonus Prompt: Launch Campaign Builder

> *Help me build a 7-day visual content launch plan for my new product drop. I want post types, caption ideas, CTA language, suggested posting times, and how to turn engagement into pre-orders. Keep it bold, cultural, and campaign-style.*

☑ AI Might Give You:
- Day 1: Hype video (Reel)
- Day 3: Founder story (Carousel)
- Day 5: Countdown + bonus offer
- Day 6: Live Q&A announcement
- Day 7: Drop post + CTA = "Let's run it up."

The Hustla's Rule:

"Pretty posts don't pay. Strategic visuals do."

Let AI help you build content that doesn't just look good, it **moves people.** Because when your brand is rolling? Your bag should be too. Let every pixel push the purpose.

Chapter Ten

Shoot Your Shot
Video Content That Cuts Through the Noise

If content is king, then video?

That's the crown.

The algorithm loves it. The people share it. And the brands that master it? They don't just sell products, they create presence.

But let's keep it real:

Creating good video content can feel overwhelming.

You think you need a ring light, a film crew, 10 edits, and a viral dance to get seen.

But nah.

All you need is clarity, consistency, and AI.

This chapter is about showing you how to use AI tools to:
- Generate scripts, hooks, captions, and hashtags
- Plan and batch record your videos like a pro
- Auto-edit, auto-caption, and even repurpose long-form into short-form
- Show up confidently and creatively, even if you're just starting

What We're Gonna Cover in This Chapter:
- How to use AI to write video scripts that hook in the first 3 seconds
- Tools that auto-edit and add captions for you (so your message lands)
- Turning one long-form video into 10 clips for every platform
- Creating reels, stories, video sales pages, and educational content
- Building confidence and staying consistent without chasing clout

Because when you shoot your shot, you don't need a thousand views, you just need the **right eyes, the right message, and the right execution.**

The Hustla's Reminder:

"People connect with motion, and motion builds momentum."

Let AI help you show up sharper, speak louder, and let your face be the brand. You ain't just selling a product, you're showing up with presence. Let's shoot.

Say It Like You Mean It: Writing Video Scripts That Hit with AI

We've all been there; staring at the camera like... "What do I even say?" You got the outfit. The lighting decent. The vibe right. But the words? **Gone.**

Video ain't just about being on camera, it's about *clarity*. If you can't get your message off in the first 3 seconds, you already lost 'em. And if you're rambling? They're swiping before you hit your point. This section shows you how to use AI to write **short, punchy, purpose-driven video scripts** that hit *before* your viewers hit that scroll.

🐌 Step 1: Understand What Makes a Script Work

Every great video script (Reel, Story, YouTube Short, ad, whatever) needs these 3 moves:

1. **The Hook** – Grabs attention FAST (first 3 seconds)
2. **The Value** – Give 'em a reason to stay
3. **The Close** – Tell 'em what to do next (CTA)

Whether it's educational, emotional, or entertaining, AI can help you write it all.

🎬 Step 2: Use AI to Generate Your Script with Ease

Prompt:

> *You are a video scriptwriter for small Black-owned brands. Write a 30-second Instagram Reel script for a Black woman-owned luxury candle business. The vibe should be soft but powerful. Hook them in, explain the benefit, and end with a clear CTA.*

☑ AI Might Deliver:
[Hook] "Your peace deserves more than a playlist."
[Value] "Our handmade soy candles are blended with mood-boosting scents that help

139

you decompress, reset, and glow from the inside out."
[Close] "Smell the shift. Tap the link and elevate your self-care game."

That's short, clear, and built for conversion.

Step 3: Remix the Same Script 3 Ways

Prompt:

> *Rewrite the same candle script 3 different ways: 1) funny and relatable, 2) poetic and soulful, 3) direct and urgent. Keep it under 60 seconds.*

AI Might Return:
Funny/Relatable:

"This is your sign to stop lighting candles from the corner store that smell like disappointment. Let's do better."

Poetic/Soulful:
"A flicker for the tired. A flame for the healers. A scent made for your rebirth."

Direct/Urgent:
"Only 20 left. Restock hits at midnight. Link in bio or miss out."
Now you're ready to record *multiple videos from one idea.*

Step 4: Create a Week's Worth of Scripts in One Sitting

Use AI to batch scripts around your themes:

Prompt:

> *Create 5 short-form video scripts for a business coach targeting first-gen Black entrepreneurs. Focus on mindset, systems, marketing tips, and funding. Each one should be under 60 seconds and include a CTA.*

AI Might Say:
- "Here's why 'just start' isn't enough"
- "3 free tools to manage your biz like a boss"
- "Grant money vs. loans: Know the difference"
- "Don't sleep on SOPs! They saved my sanity"

- "DM me 'funded' for my free resource list"

Now you've got a full **video series** ready to shoot on your terms.

Real-Life Example: Chris the Mobile Mechanic

Chris runs a mobile auto repair business in Atlanta. He's not flashy, but he's real. He started posting short videos breaking down common car issues.

AI helps him:
- Write his intros
- Create analogies that make sense
- Add CTAs like "Text me the word OIL for a discount this week"

Result?

400% increase in DMs

Landed a corporate fleet maintenance contract

People say: "You made me trust you before I even booked you"

All because his message was clear, and his script was solid.

Bonus Prompt: Shoot-Ready Script Kit

Write 7 video scripts for a nonprofit helping formerly incarcerated Black men get job training. Include tone (empowering, no pity), hooks that speak to justice and opportunity, and a CTA to donate or share.

Now you're not just telling a story, you're starting a movement.

The Hustla's Rule:

"Don't freestyle your future."

AI helps you script content that's sharp, confident, and clear so your message lands every time.

Because when you speak with purpose? **The people, and the money listen.**

Edit Like a Pro Without a Crew: AI Tools to Polish Your Videos

So you finally shot the content. The lightings alright. The message is solid. You even hit the hook in the first 3 seconds. Now what? If you're like most hustlas, that raw video footage just sits in your phone for days, maybe weeks, because **editing feels like a whole other job**.

You either don't know how to make it look clean, you don't have time, or you're not trying to spend $200 on Adobe Premiere and 3 hours on YouTube tutorials.
But here's the blessing:

AI took the editing process and simplified the whole game.

You don't need to be a filmmaker.

You just need a system that lets you:
- Trim
- Add captions
- Insert music
- Layer in CTAs
- Format it for IG, Reels, YouTube Shorts, and even your website

Let's walk you through exactly how to do all that with zero stress and max results.

The Core AI Video Editing Stack

These are your go-to tools:

1. Descript (Best for script-based editing + transcription)
- Turns your video into a Word-style document
- Delete text to delete video sections
- Automatically removes "ums," "uhs," and awkward pauses
- Auto-generates captions and formats them
- Great for interviews, tutorials, and content where your *words* matter

2. CapCut (Best for short-form, mobile edits)
- Built-in templates for Reels, Shorts, and TikTok
- Auto-captions with stylish fonts and effects
- Sound syncing with music and voice
- Easy drag-and-drop interface with transitions
- Mobile AND desktop versions available

3. Runway ML (Advanced edits & special effects)

- Remove backgrounds without green screens
- Add cinematic transitions and filters
- Smart audio cleaning and visual enhancement
- Good for branded product visuals or higher-end storytelling

4. Davinci Resolve (When you're ready to level up)

- Full video editing suite with color grading, motion graphics, and sound mastering
- Best used on desktop for YouTube-level production
- Takes longer to learn, but the results are **Netflix-worthy**

Your 7-Step AI Editing Flow (Step-by-Step)

Let's say you just recorded a 60-second tip video for IG Reels. Here's how to flip it into polished content in under 60 minutes:

Step 1: Upload to Descript

- Import your video file (.mp4 or .mov)
- Descript automatically transcribes your words
- Read your words like a doc; delete anything you don't want
- Use "Filler Word Removal" to clean up your speech

Pro Tip: Highlight your key phrases so you can turn those into **on-screen captions** in Step 3.

Step 2: Export Clean Version to CapCut

Once Descript trims your message, export the cleaned video and upload it to **CapCut** (mobile or desktop).

Here's what you'll do:
- Drop in **captions** using CapCut's auto-caption tool
- Choose your font, color, and animation style
- Add your **logo in the corner** (Canva exports work perfect)
- Drop a **soundtrack** or ambient beat behind your message
- Insert **B-roll** or overlay clips (like product footage or photos)

Pro Tip: CapCut templates can save you HOURS. Search for "Motivational" or "Product Demo" and pick one that fits your brand.

Step 3: Add CTAs and Visual Punch

Now that the core video is clean, finish strong with a CTA your audience can feel.

Prompt for AI:

> *Give me 5 call-to-action captions I can use on Reels that speak to Black entrepreneurs, encouraging them to DM, click, or comment without sounding like a sales pitch.*

Examples:
- "Ready to level up? Let's build. DM me 'FUNDING.'"
- "Drop a   if this hit and tag someone who needs this."
- "Tap the link. You already behind if you wait."
- "This is your reminder. Don't just dream. MOVE."
- "Let's talk solutions. DM me your biggest challenge."

Use CapCut to add these lines as **animated text overlays** near the end of your video. Keep it bold, brief, and branded.

Step 4: Format for Every Platform

Let AI or CapCut help you reformat the same video for:
- IG Reels (9:16 vertical)
- YouTube Shorts (same size, diff platform)
- Facebook (1:1 square works best)
- Website hero sections (16:9 landscape)

This way, **one piece of video gets you visibility in five places** without filming again.

Step 5: Schedule with Captions + Hashtags

Don't stop at editing. **Get it scheduled.**

Prompt AI:

Write a caption and hashtag set for this video. I'm targeting Black creatives trying to build personal brands. Keep it motivating, not corny. Include 7–10 hashtags that blend culture and entrepreneurship.

Example caption:
"The world don't need another perfect brand. It needs a real one. Start with what you got. Use what you know. Just move."

#BlackBusiness #CreativeHustle #ContentThatHits #BuildYourLegacy #ReelReady #AuthenticOverPerfect

Real-Life Example: DJ Rob from Chicago

DJ Rob wanted to grow his music academy but didn't know how to market on video. He used to post low-quality IG lives with no captions and no structure.

Then he ran this exact flow:
- Used ChatGPT to script 10 tips for beginner DJs
- Recorded 10 short clips in one hour
- Uploaded to Descript → CapCut
- Added auto-captions, a beat behind each one, and CTA overlays
- Scheduled all of them across IG Reels and YouTube Shorts

Booked 4 new clients in 2 weeks

Got asked to speak at a music conference

Started a monthly beat sale from DMs that came from his videos

Bonus Prompt: AI-Backed Video Editor Assistant

Build me a 6-step workflow for turning raw iPhone footage into a polished, branded video using CapCut and Descript. Include when to add captions, when to trim filler, and how to format for both IG Reels and Facebook.

AI Returns:
1. Import raw video into Descript
2. Trim stutters, filler, and add auto-captions
3. Export trimmed file to CapCut

4. Add text overlays (CTAs, branding)
5. Insert music and B-roll if needed
6. Export 9:16 for IG, then 1:1 for Facebook

The Hustla's Rule:

"You don't need a production budget; you need a process."
AI turns your camera roll into a marketing machine.

Because when your edits are tight and your message is clear? **That content don't just exist, it performs.**

Run It Back: Repurposing Your Videos Like a Boss

Creating content is one thing but **multiplying it?** That's where the bag really lives. You ever notice how the smartest brands post the same message *10 different ways*? It ain't because they ran outta ideas, it's because they built a system.

They film one video...

Then stretch that message into **Reels, carousels, tweets, emails, YouTube Shorts, audiograms, and blog posts.**

And now with AI?

You can do the *exact same thing,* even if you recorded it on your phone and edited it in CapCut while eating leftovers.

Let's break this down.

Step 1: Feed AI the Original Video Transcript

Start by transcribing your video using:
- **Descript** (automatic transcription)
- **ChatGPT's Whisper plug-in**
- **CapCut's auto-caption feature**

Once you've got the text from your video, feed it to AI and say:

Prompt:

This is the transcript from my 90-second video. Help me repurpose this content into 7 new pieces of content across Instagram, Facebook, email, and YouTube. My tone is motivational, urban professional, and culture-rooted.

AI Might Return:
- A short motivational quote (for an IG Story)
- A tweet thread breaking down your 3 core points
- A carousel post: "What Nobody Tells You About [Topic]"
- A Reel idea with a new visual angle
- An email to your list: "This Hit Me This Week…"
- A YouTube Shorts title: "Here's Why You're Still Burnt Out"
- A Pinterest or LinkedIn graphic suggestion

That's **one video → seven opportunities.** Minimum.

Step 2: Slice and Dice with Descript or CapCut

Take your full video and break it into 15- to 30-second clips.

Clip types to extract:
- The hook
- The main takeaway
- The emotional punch
- A relatable "real talk" moment
- The CTA
- The story segment

Prompt for AI:

Read this transcript and tell me which 3 moments I should cut out and post as standalone micro-videos. Format the timestamps for CapCut editing.

Pro Tip: These smaller clips make great Reels, Facebook Shorts, or even TikTok if that's your lane.

Step 3: Turn Video into an Email or Newsletter

Don't waste that story, **expand it.**

Prompt:

> *Turn this 60-second video into an email to my subscribers. Break down the lesson,*
> *add a personal intro, and end with a CTA to check out my new service offer.*

Now you've got an email with:
- A subject line
- An opening hook
- A reflection or value piece
- A soft call to action

Post the video on IG, then follow up via email the next day. That's **omnichannel marketing.**

Step 4: Caption It for Every Platform

Different platforms = different energy.

Let AI remix your video into **platform-specific captions.**

Prompt:

> *Write 4 captions for this 30-second video. One for IG Reels, one for Facebook, one for*
> *YouTube Shorts, and one for Threads. Keep the message consistent but tailor the*
> *tone and CTA for each platform.*

AI Delivers:
- **IG Reels:** "If you've been waiting on a sign, this is it. Don't scroll, apply."
- **Facebook:** "We gotta talk about how burnout becomes the norm for entrepreneurs."
- **YouTube Shorts:** "Stop playing with your purpose. Hit that subscribe for real talk every week."
- **Threads:** "Legacy over hype. Every time."

Real-Life Example: Sam the Speaker

Sam speaks at schools, churches, and virtual summits. He recorded one 3-minute keynote clip about mindset and uploaded it to YouTube.

Using AI, he:
- Created 10 social clips from that one speech
- Built a week's worth of posts using key quotes + transcripts
- Got captions for LinkedIn, IG, and his email list
- Used ChatGPT to write a blog post: "What I Wish I Knew Before My First Speaking Gig"

One video turned into:
- Content for 2 weeks
- 3 new bookings
- A DM from a podcast with 40K followers

All off one message, repurposed with **intention.**

Bonus Prompt: Repurposing Flow Builder

> *Create a repurposing workflow for my weekly 60-second IG video. I want to post it on IG Reels, Facebook, YouTube Shorts, email, and Twitter. Include editing tools, post types, AI prompt suggestions, and a weekly schedule to follow.*

AI Might Build:
- **Monday:** Edit in CapCut → post to Reels
- **Tuesday:** Chop 15-sec clip for FB Shorts
- **Wednesday:** Tweet a quote + link to Reels
- **Thursday:** Email blast using transcript & AI copy
- **Friday:** Post a recap carousel on IG with CTA to replay

Now you're building content like a whole creative agency with just your voice and a vision.

The Hustla's Rule:

"Don't just create content, create assets"
.

One video can live 10 lives if you know how to flip it. Let AI help you multiply your message, so your name stays in the feed, the inbox, and the conversation.

Less content. More impact. Run it back.

Close with Confidence: Using Video to Sell Without Feeling Salesy

Let's be real, most folks out here got **views with no value**.

They're getting likes, but not bookings.
They're getting shares, but not sales.
They're talking loud, but the bag still ain't moving.

Why? Because they're missing the bridge between content and conversion.
That's what this section is about:

Using video content to SELL without feeling like a used car ad.

Letting your videos work like digital salespeople, guiding your audience from interest
→ trust → *action*.

Step 1: Build a Simple Video Sales Funnel

Every video should serve a role in your sales journey. Here's a basic 3-part funnel:

1. Awareness Video
Show up with something scroll-stopping, emotional, or relatable.

Example:
"3 reasons I had to quit my job and bet on myself."
Hook them with your story or a relatable pain point.

2. Value Video
Teach, demonstrate, or break something down.

Example:
"Here's how I built my business using just 2 free tools."
This shows your credibility *without* begging for attention.

3. Offer Video
Direct, clear, and confident.

Example:
"If you're tired of doing this alone, I got a program that breaks it all down. Doors close
Friday. Tap the link."

This is where the money comes in.

Pro Tip: Pin all 3 videos to your IG Reels or drop them as a carousel on Facebook.

Step 2: Let AI Write Video Sales Scripts that *Don't* Sound Corny

Prompt:

> Write a 60-second video script for my grant-writing service. Audience: Black women entrepreneurs who've been rejected before. Keep it real, empowering, and end with a CTA that invites them to book a call, not "buy now" energy.

AI Might Return:
"You've been told 'no' by people who never took the time to understand your vision. But that don't mean your business ain't fundable. I've helped dozens of women get the funding they were told they'd never see. You ready for your yes? Hit the link!

Let's work."

That's **storytelling that sells.**

Step 3: Use Testimonials That Tell, Not Just Show

People don't buy products, they buy **proof.**

Instead of just saying, "This helped me," train your clients to drop **testimonial-style video content** that tells a story.

Prompt:

> Write a short script for a client to follow when recording a testimonial video about my web design service. Make it easy, authentic, and focused on what changed after we worked together.

Script:
1. "Before I found [your brand], I was struggling with…"
2. "What made me trust them was…"
3. "Since launching my site, here's what happened…"
4. "If you're on the fence, don't wait. Just do it."

Drop these videos into your sales page, IG Highlights, or reels with a "Client Love" label.

Step 4: Pair Video with Lead Magnets or Booking Links

Don't post and ghost.

Every sales-focused video should lead somewhere; a calendar, an opt-in, a sales page, a DM convo, something.

Prompt:

> *Write a caption and CTA for this 30-second Reel promoting my financial coaching service. I want people to feel empowered, not pressured. The CTA should lead to my free guide + booking link.*

Example Caption:
"You're not bad with money. You were never taught. This video might be the mindset shift you've needed. Grab the free guide, then book the call. It's time to take the power back."

Real-Life Example: Rochelle the Resume Coach

Rochelle was tired of cold selling in DMs.

She wanted her video content to *do the work*.

Here's her flow:
- Drops a value-based video tip every Monday ("How to answer tricky interview questions")
- CTA at the end: "Want the full guide? Link in bio."
- Her free guide auto-collects emails and invites viewers to a strategy call
- After call, she closes 1 in every 4 leads

She used AI to write the video scripts

CapCut to polish the video

ChatGPT to write email follow-ups and webinar invites

All from her living room with a $20 tripod

And now she doesn't chase leads.

She attracts them.

Bonus Prompt: Launch Series Builder

Build me a 5-part video launch sequence to promote my upcoming merch drop. Include: 1 hype video, 1 behind-the-scenes clip, 1 value-based piece, 1 FAQ video, and 1 final push before the launch ends.

AI might return:
- Video 1: "Why This Drop Matters to the Culture" (Hype)
- Video 2: "Come Behind the Scenes with Me" (Trust)
- Video 3: "3 Reasons We Chose This Design" (Value)
- Video 4: "Here's What Y'all Been Asking" (Engagement)
- Video 5: "Last Day to Cop. Don't Miss Out" (Urgency)

Now your videos don't just look good, they're **strategic, intentional, and built to move product.**

The Hustla's Rule:

"If your content ain't converting, it's just entertainment."

Let AI help you close with clarity, confidence, and *culture.* Because when your content moves hearts **and** pockets, you don't just shoot your shot.

You score.

Chapter Eleven

C.R.E.A.M.
AI Gets the Money

You built the brand.

You got the content.

You been consistent on socials, got folks in your DMs, and your site finally movin' product.

Now here's the real question:

Do you actually know where your money's going?

Because momentum means nothing if the money's leaking. Most hustlas are making moves but not managing. Spending heavy. Tracking light. Mixing business and personal. And when tax time comes?   We ducking instead of deducting.

This chapter is all about **using AI to boss up your finances.**

We're gonna help you:
- Track your income + expenses
- Automate your bookkeeping
- Forecast your next quarter
- Understand your numbers *without* needing a CPA degree
- And use those insights to make real money moves

What You'll Learn in This Chapter:

- How to use AI to build a budget, manage cash flow, and analyze profit
- Tools that help you automate invoices, receipts, and reports
- How to use AI to create financial dashboards and prep for tax season
- Using AI to plan for growth, investments, and long-term legacy

Because financial freedom ain't about making money. It's about **keeping it, flipping it, and understanding it.**

The Hustla's Reminder:

"Cash rules, but clarity scales."

Let AI handle the math while you focus on the mission. This ain't just about income, it's about ownership. Let's run the numbers.

Count Every Coin: Tracking Income & Expenses with AI

You can't scale what you don't track. And too many of us are out here making money but can't tell you how much we *actually* net. You get paid from Cash App, Stripe, Shopify, Zelle, PayPal, and maybe even cash on hand...

But when it's time to break it down?
    You guessing.
    You scrolling through screenshots.
    You praying your tax person don't judge you too hard.

Nah. That ain't the move.

This section is all about using AI to **track your income, manage your expenses, and see the full financial picture** clearly, cleanly, and consistently.

Step 1: Choose Your Tools

You don't need QuickBooks (unless you're ready for it).

You just need tools that talk to each other *and* to AI.

Starter Toolkit:
- **Google Sheets** – Create a simple income/expense tracker
- **ChatGPT or Claude** – To help you clean and analyze your data
- **Wave Apps** – Free bookkeeping with receipt scanning & invoicing
- **Notion + Tally** – For customized tracking dashboards
- **Zapier** – Automate money data from payment platforms to your spreadsheet

Step 2: Build a Basic AI-Friendly Tracker

Let's build your spreadsheet like this:

Website Sale – T-shirt	$0	Sales	Shopify

Prompt:

You are a financial assistant. Help me set up a basic income/expense tracker in Google Sheets. I want columns for date, description, income, expense, category, and notes. Also suggest how to automate recurring expenses and calculate monthly profit.

AI will:
- Create your formulas
- Suggest drop-downs for categories
- Help you calculate net income for the month
- Recommend color-coding to highlight profit vs. loss

Step 3: Automate the Input

Instead of manually typing in numbers every day:

Use Zapier or Make.com to automate these moves:
- When a Stripe payment hits → log it to Google Sheets
- When a Shopify sale comes in → create a new entry
- When a recurring charge hits your business account → auto-tag it as "Subscription" in your sheet

Prompt:

Help me create a Zapier automation that takes new sales from PayPal and adds them to my income tracker in Google Sheets. Also categorize them based on product type (coaching, merch, digital downloads).

Step 4: Let AI Spot Patterns and Give Feedback
Once you've got a few weeks of data, let AI tell you what's up.

Prompt:

> *Based on my income and expense tracker [paste data], analyze where I'm spending the most, what my most profitable product is, and give me 3 tips to improve cash flow.*

AI might return:
- "You're spending 35% of income on software. Consider downgrading subscriptions."
- "Your merch is profitable but selling inconsistently. Try batching promo around paydays."
- "You've made $2,400 in the last 30 days but only netted $1,200. Time to tighten expenses or raise pricing."

Now you're not just tracking, you're making *decisions.*

Real-Life Example: Cedric the Chef

Cedric runs a meal prep business and uses to do his books off memory and Cash App screenshots.

Once he used AI:
- He set up a Google Sheet tracker
- Had ChatGPT generate formulas to show profit by week
- Used Zapier to auto-log his Stripe payments
- Asked AI weekly: "Where did I spend too much?"

In 60 days, Cedric cut his expenses by 22%, started saving $500/month, and was finally ready to apply for a business line of credit.

Bonus Prompt: Financial Clarity Snapshot

Based on this month's income and expense data, write me a summary report. Include total income, total expenses, biggest cost category, profit margin, and 3 smart money moves I should make next month.

This becomes your **monthly CEO check-in.** No spreadsheets? No problem. AI breaks it down in plain English.

The Hustla's Rule:

"If you ain't tracking the money, you ain't building wealth. You just touching cash."

Let AI help you count every dollar, trim the fat, and move with clarity.

Because real freedom ain't in the flex, it's in the **financial facts.**

Set It and Stack It: Automating Your Financial Systems with AI

You got the hustle.

You got the sales.

But if you're still sending invoices manually, forgetting to log expenses, or pulling up screenshots when it's time to do your taxes?

That's stress you don't need.

The goal is freedom. And freedom comes from **systems**.

This section is about building financial automations so smooth, it feels like you got a digital accountant on payroll. Except… it's AI, and it don't sleep, complain, or invoice you at the end of the month.

Step 1: Choose the Tools That Talk to Each Other

You don't need to drown in software. You just need your apps to connect.

Money Stack That Works:

Receip Scannin	Wave / Expensify	Snap and auto-log
— ..		..
AI Insigl	ChatGPT	Analyzes, recommends, summarizes

Step 2: Auto-Generate and Send Invoices with AI

Prompt:

> *Create a branded invoice template for a Black-owned consulting business. Include name, client info, service description, price, payment methods, late fee clause, and thank-you note.*

AI Delivers:
- Clean, professional copy
- Terms that sound like you, not like a lawyer robot
- Can be copied straight into Wave or Canva for visual design

Automate It:
- Set up recurring invoices for monthly clients
- Use Wave or HoneyBook to send, remind, and collect automatically
- Set late fee reminders at 5 and 10 days with auto-email follow-up

You don't chase payments. The system does.

Step 3: Scan Receipts & Log Expenses on Autopilot

Use Tools Like:
- Wave's mobile app
- Expensify
- QuickBooks Self-Employed
- Google Lens → Google Drive folder → Auto-import to Sheets

Zapier Flow Example:

When a photo of a receipt is added to Google Drive, auto-send it to my expense tracker with today's date and category suggestion.

Prompt for AI:

> *I uploaded 10 receipts from the last 2 weeks. Can you log the amounts and suggest categories like meals, tools, subscriptions, and travel?*

AI turns a pile of screenshots into *organized tax deductions*.

Step 4: Let AI Do Weekly or Monthly Financial Reviews

You ain't gotta sit with a calculator for 3 hours. Let AI tell you what your books are saying.

Prompt:

> Review my income and expense log [paste data or summary]. Write me a report like I'm a CEO who wants to know what's working, what's leaking, and what I need to change.

AI Might Return:
- "Your top client brought in 42% of your income. Time to secure that relationship."
- "Subscriptions are creeping up; you've got 4 tools doing the same thing."
- "Ad spend has ROI, but your return is dropping. Consider testing other platforms."

Let AI write this like a monthly money check-in for your journal or business meeting.

Step 5: Build a Notion or Airtable Money Dashboard

Let AI help you design a live dashboard where you can:
- View profit/loss
- See income by product
- Track recurring expenses
- Visualize your financial growth

Prompt:

> Build me a Notion dashboard for tracking business income and expenses. Include income streams, recurring charges, goal tracker, and monthly profit visuals.

AI might sketch a board like:

Stream	April	May	Goal

Now your money is visual. Real. Motivating.

Real-Life Example: Layla the Lash Artist

Layla runs a booming beauty brand but was drowning in manual work writing invoices in Notes, tracking appointments on paper, and guessing her profits.

She used AI + automation to:
- Create invoice templates with ChatGPT
- Auto-send and track them using Square
- Scan receipts into Wave
- Let AI summarize her income vs. product cost every month

Now she:

Saves 6 hours a week

Sends invoices before clients leave the chair

Raised her prices based on real profit margins

Automation unlocked *CEO energy*.

Bonus Prompt: Build My Financial Automation Game Plan

> *Build a step-by-step automation strategy for a solo service provider who gets paid via PayPal, Cash App, and Zelle. I want to auto-log income, track expenses, send weekly invoices, and create a monthly summary I can review every Sunday.*

AI Returns:
- Create Google Sheet with tracking formulas
- Use Zapier to log PayPal + Stripe into Sheet
- Scan and tag receipts with Expensify
- AI-generated monthly reports sent to email
- Sunday calendar reminder: "Money Check-In with ChatGPT"

The Hustla's Rule:

"The less time you spend chasing paper, the more you can focus on multiplying it."
Let AI and automation handle the routine, so you can handle the vision. Because real bosses don't touch every dollar, they build systems that do it for them.

See It Before You Spend It: Budgeting & Forecasting with AI

Let's be honest:

Most hustlas treat budgeting like a punishment. Too many of us are allergic to planning money, because the game taught us to move fast and fix later.

But hear this: **money without a map don't multiply.**

You can't build wealth on vibes. You gotta build it on **vision and numbers.**

This section is about how to use AI to:
- Build realistic budgets
- Predict your income
- Spot slow months before they come
- And make sure you're not just surviving, but *scaling*

Step 1: Build Your "No Stress" Budget Framework

AI can help you build a simple budget based on your real goals, not some generic template.

Prompt:

> *You are my financial planner. Help me create a monthly business budget based on these averages: $5K income, $1.2K software/tools, $800 marketing, $1.5K product/materials, and $1K to pay myself. Make sure I have room for taxes, emergency stash, and future investments. Format it as a table with suggested percentages.*

AI Returns:

Categor	Amount	% of Income
Owner Pa	$1,000	20%

AI can tweak it by goal: "Build savings faster" or "Scale paid ads."

Step 2: Forecast Next Month Based on This Month

Instead of guessing what your next move should look like, AI can help you **analyze your trends and predict your income + expenses.**

Prompt:

> *Based on my last 3 months of income and expenses [paste sample data], forecast next month. Include expected sales, top 3 expense categories, and any patterns I should know about.*

AI Might Say:
- "You average 15% growth month-over-month. Expect $5,750 next month"
- "Product costs spike during launch weeks; budget accordingly"
- "Client work dips during holidays, plan a digital promo instead"

That's *strategy*. You're not reacting you're *ready*.

Step 3: Set Revenue Goals You Can Actually Hit

A goal without a plan is a wish. AI helps you map the math behind your mission.

Prompt:

> *Help me set a 3-month revenue goal. I sell coaching ($1,000), digital downloads ($49), and merch ($35). I want to make $10K/month. Show me how many of each I need to sell per week to get there.*

AI Delivers:

Merch	$35	100	25

Then AI might say: "Focus on bundling merch + downloads and promote coaching on high-engagement weeks."

Now your goals ain't just big, they're **breakable and trackable.**

Step 4: Build a Cash Flow Plan with Cushion

Revenue looks good, but what about your **cash flow?** That's the difference between "I made 5K" and "I got 5K *available.*"

AI can help you plan for:
- Delayed payments
- Large drops (bulk orders, restocks)
- Slow seasons
- Emergency funds

Prompt:

> *Build me a 90-day cash flow plan for a small e-com business that brings in $6K/month but spends heavily every 3rd month on restocks. I want to make sure I don't go broke during slow weeks.*

AI Returns:
- Breakdown by week
- Flags restock months
- Suggest putting 20% aside during high weeks
- Build in a 2-month cushion for operating costs

Real-Life Example: Marlon the Mobile Barber

Marlon's business booms during holidays but slumps every January. Before AI, he'd panic in the slow months and overspend during the hot ones.

Now he:
- Uses AI to review year-over-year income trends
- Forecasts seasonal dips and preps "Holiday Hustle Packs" ahead of time
- Built a 12-month spreadsheet with AI that predicts income based on bookings + product sales

Result: No more overdrafts. No more guesswork.
He even started saving for his mobile barbershop truck 2 years ahead of schedule.

Bonus Prompt: Hustla Budget & Forecast Dashboard

Build me a Notion or Google Sheet-based system to budget, track income, forecast earnings, and set monthly financial goals. Include formulas and suggestions for automating with Zapier or AI assistants.

AI Delivers:
- Monthly summary
- Income stream breakdown
- Expense tracker
- Profit goals
- Forecast vs. Actual comparison

The Hustla's Rule:

"Money loves direction."

Let AI help you move like a CFO, not just a content creator. Because when you start forecasting like a boss? Your income don't just show up, it shows *out*.

Beat the Deadline: AI for Tax Prep & Financial Reporting

Let's be honest, tax season got some of us shook.
It's not that we didn't make money.

It's that we don't have our *receipts, reports, or records* ready when the IRS come knocking.

Scrambling to find expenses

Guessing how much we owe

Forgetting deductions

Mixing business with personal accounts (you know who you are   )

But it don't have to be like that.

AI can help you get ready early, move smarter, and dodge those tax-season panic attacks.

This section is your step-by-step breakdown for using AI to:
- Get your numbers together
- Separate business and personal

- Prepare reports for your tax pro
- Identify deductions
- And stay 10 steps ahead of the IRS

Step 1: Start Organizing NOW (Not in March)

AI can't save you from procrastination. But it can save you from confusion.

Prompt:

> *Build me a checklist of everything I need to get ready for tax season as a sole proprietor who sells products and services online. Include what reports to pull, what tools to use, and what to give to my tax preparer.*

AI Might Return:
- Income reports from Stripe, PayPal, Cash App
- Expense breakdown (by category)
- Receipts for equipment, marketing, software
- Mileage log (if you drive for business)
- Home office calculation (if you qualify)
- 1099-K forms from platforms like Etsy, Shopify, etc.
- W-9s from any subcontractors you paid over $600

Bonus Tip: Create folders in Google Drive labeled "2024 Taxes" → Income / Expenses / Receipts / Forms

Step 2: Use AI to Categorize and Summarize Everything

If you've got a spreadsheet with your income and expenses, AI can analyze it fast and break it down like a pro accountant.

Prompt:

> *You are a tax assistant. Here is my business income and expense data [paste it]. Organize my expenses into tax-deductible categories (meals, travel, marketing, home office, etc.) and summarize total income, total expenses, and net profit.*

AI Gives You:
- Total income

- Total expenses
- Estimated net profit
- Expense breakdown by type
- Suggestions for missing deductions

Now you can hand that to your CPA and say: "Here's the file."

Step 3: Estimate Taxes Ahead of Time (No Surprises)

Don't let tax time sneak up on you.

Prompt:

> *Estimate how much I should set aside for taxes based on this income and expense summary. I'm a single-member LLC with no employees, based in Georgia. Assume I want to stay safe and not owe anything at the end.*

AI Might Say:
- "You should set aside roughly 25–30% of net income."
- "That includes self-employment tax, federal, and Georgia state."
- "If you earned $60K net, aim to save $15K for taxes."

Bonus: Use apps like **QuickBooks Self-Employed, Wave**, or **Keeper** to automate tax withholding.

Step 4: Use AI to Write Financial Reports for Grant Applications or Funding

If you ever apply for:
- Business grants
- Lines of credit
- Community funding
- Pitch competitions

They're gonna want to see your **financials.**

Prompt:

Write a 1-page financial summary for my business based on this income and expense data. Include key growth stats, revenue sources, and how I plan to scale. Keep it simple and confident like a founder that's clear and focused.

Result: Clean, powerful, funder-friendly report you can drop into a deck or application.

Real-Life Example: Keisha the Clothing Brand Owner

Keisha was hit with a $3,500 tax bill last year because she didn't know about write-offs or quarterly taxes.

This year, she:
- Used AI to write a deduction list and identify gaps
- Built a monthly report template in Notion with ChatGPT's help
- Used Keeper Tax to scan her bank account for deductions
- Had AI draft a clean P&L (Profit & Loss) report for her accountant

Now?

She knows her numbers monthly

She's setting aside tax money automatically

And she walked into her tax prep meeting like a CFO, not a confused creator

Bonus Prompt: Full Tax Prep System Builder

Build me a full tax prep system for my digital product business. Include monthly tracking, quarterly tax payment reminders, report templates, and a checklist for year-end. I want to automate as much as possible and keep it stress-free.

AI Might Deliver:
- Monthly Google Sheet log
- Quarterly calendar reminders
- Email draft to send to CPA with all files
- Tax folder structure for Google Drive
- Estimated tax calculator template

Now you're not just getting ready, you're staying ready.

The Hustla's Rule:

"Tax season don't have to be stressful, it just gotta be strategic."

AI gives you peace of mind, organized receipts, clean reports, and a paper trail that protects your grind. Because if you plan ahead, you don't just survive tax season; you **own it.**

Chapter Twelve

All Eyez on AI
The Blueprint for Success

Let's pause for a second and look at how far you've come. You started this journey maybe feeling overwhelmed by AI, unsure if it could really help *you*, unsure if it was built for us, for the culture, for the hustle, or for the come-up.

But now?

You ain't just using AI; you're **commanding it**.

You've learned how to:
- Write prompts that produce power moves
- Use AI to market, design, schedule, and scale
- Automate tasks that once drained your time
- Protect your brand, run your books, and forecast your future
- Build a digital infrastructure around your purpose

That's legacy in motion.

This final chapter is all about:
- Pulling it all together
- Setting a vision for what's next
- And giving you the blueprint to run this play forever

Because AI is not the finish line, it's **the foundation**.

What You'll Lock In With This Chapter:
- What your business should look like now (if you've really been using these tools right)
- A full blueprint for daily, weekly, and monthly AI-powered success
- Mindset shifts to protect your peace as you grow
- A plan to scale without burnout, confusion, or chaos
- The confidence to lead like a boss in an automated, accelerated world

The Hustla's Reminder:

"You didn't just learn how to use AI; you learned how to use it like a visionary."

This ain't about staying busy. This is about building something that lasts. You put in the work. Now let's map out how to *keep winning with it*. All eyez on AI. Let's finish strong.

The Hustla Rebuild: What Your Business Should Look Like Now

If you've been rocking with this book chapter after chapter applying these prompts, building these systems, and shifting how you move, then by now, your business should be *unrecognizable* (in the best way possible).

You're not just running your business.

You're **leading it with vision, clarity, and tools that scale.**

This section is about **what should be true** in your operation if AI is now your daily partner. Think of this as your *business glow-up checklist*.

Your Digital Infrastructure is Solid
You've got systems in place, not just vibes.
Your content calendar is planned, branded, and batch-created ahead of time.
You're using tools like ChatGPT, Canva, Descript, and CapCut like second nature.
You're not logging into 10 apps, you're using integrations and automation to reduce the noise.
Your website, social, and brand voice are **aligned and consistent**.
You're no longer hustling for every dollar; you're building a **brand machine.**

Your Money is Making Sense
You know your monthly income, expenses, and profit *before* the 30th hits.
You've got an AI-powered spreadsheet or Notion dashboard that tracks it all.
You've automated invoicing, receipt capture, and even tax prep.
You're forecasting your goals, not just praying for a good month.
Your pricing reflects your value and your confidence.
Bottom line: You're not just touching money. **You're managing money.**

Your Marketing Feels Like Movement

You're not waking up asking "What should I post today?"

Your scripts are written, captions are scheduled, and Reels are auto-edited.

You've created funnels that sell *while you sleep*.

Your brand voice is locked in.

You're repurposing content like a creative agency.

Your audience isn't guessing who you are, they *feel* it every time you post.

And more importantly, your people are *engaged*. Because your content speaks **to them, for them, and with them.**

Your Mindset is CEO-Level

You're working *on* the business, not just in it.

You've stopped trying to do everything yourself and started building systems that support your energy.

You've reclaimed your time, your focus, and your creativity.

You don't fear AI; you *leverage* it.

You're not reactive; you're strategic.

Your moves are intentional. Your goals are structured. And your brand now has *a presence, a process, and a purpose.*

Real-Life Example: You.

That's right, **you.**

If you've applied even 30% of the sauce in this book, you should already:

- Feel less overwhelmed
- Be showing up more consistently
- Have smarter systems
- Be getting more engagement and inquiries
- Feel like you've got *a partner in AI* and not just a tool

If not? Don't trip. It's not about being perfect, it's about being *active*. Run it back. Revisit the prompts. Start slow, build momentum. You've got the blueprint now.

Bonus Prompt: Hustla Self-Audit

Write me a one-page self-assessment to check where I'm winning and where I need to tighten up. Focus on systems, content, finances, and mindset. Be honest but motivating; like a coach that's rooting for me.

AI Might Say:
- "You've built strong visual systems, but your financial flow needs clarity."
- "Content is hitting, but repurposing isn't fully optimized yet."
- "Your mindset has shifted, now let's build the habits to match."
- "You're one system away from full-time freedom."

The Hustla's Rule:

"Once your systems catch up to your vision, you become unstoppable."

AI ain't just a cheat code. It's the co-founder you didn't know you needed. Let this be the moment where your business stops surviving and starts *scaling on purpose.*

The Hustla Blueprint: Your Daily, Weekly, & Monthly AI Game Plan

Success ain't about doing more, it's about doing what matters **consistently.**

And now that AI is part of your business, it's time to treat it like what it is:

Your **Digital COO** handling your marketing, your money, your moves, and your mental.

This section is your **TrapGPT Hustla Schedule,** a repeatable rhythm you can run every day, week, and month to stay sharp, stay scalable, and stay in your bag.

DAILY: *Power Moves, Not Panic*

1. Morning Command Prompt
Start the day with your Digital Assistant.

Prompt:

What are 3 tasks I should focus on today to grow my brand, connect with my audience, or generate sales?

2. Engage with Purpose
Check your IG/FB comments and DMs. Use AI to help you reply in your brand tone.

Prompt:

> *Help me respond to these 3 DMs in a way that's professional but personal. One's about pricing, one's about availability, and one's about a collab.*

3. Post with Precision
Use your pre-written AI content plan (from ChatGPT or Lately) and post one Reel, carousel, or quote card.

4. Track Sales or Engagement Briefly
Check performance. What's getting clicks, saves, and shares?
Ask AI:

What's one thing I should post tomorrow based on how today's content performed?

WEEKLY: *System Over Stress*
Sunday Hustla Reset (60–90 min)

1. Plan Content with AI
Ask AI for next week's content based on your goals.

Prompt:

> *Plan a week of IG posts for my coaching brand. One offer post, one tip, one story, one testimonial, and one CTA.*

2. Batch Create & Design
Write captions, design Canva graphics, and script any short-form videos.

3. Schedule Posts
Load everything into your scheduler (Lately, Metricool, Meta, etc.)

4. Check the Bag
Open your income/expense tracker (Google Sheets, Notion).
Ask AI to summarize it like a CEO check-in.

Prompt:

Based on this week's income and expenses, what's my profit, and what should I look out for next week?

5. Set the Tone for the Week
Ask AI to give you a weekly mindset note, journal prompt, or quote.

Prompt:

I'm a Black entrepreneur trying to stay grounded and focused this week. Give me a quote or reminder that'll keep me locked in.

MONTHLY: *CEO Energy Only*

1. Run the Numbers
Ask AI for your monthly breakdown:
- Top income streams
- Highest expenses
- Net profit
- Recommendations for cuts or investments

Prompt:

Create a monthly business report from this spreadsheet [paste summary]. Include total revenue, expenses, profit, growth percentage, and 3 things I can do to increase profit next month.

2. Review Brand Metrics
Use AI to evaluate your content strategy.

Prompt:

Which of my last 10 posts had the best performance and why? What topics or styles should I focus on more next month?

3. Set Next Month's Goals
Let AI build a plan around your numbers and goals.

Prompt:

Help me plan next month's goals. I want to grow my email list, sell 20 digital products, and increase my IG engagement. Give me a breakdown with weekly focus points.

4. **Create a Monthly Blueprint PDF**

Turn all of that into a one-pager you review or send to your accountability partner/investor/coach/team.

Real-Life Example: Trina the Techpreneur

Trina runs a digital marketing business from Atlanta. She was doing everything in real time and feeling overwhelmed. Since implementing her AI game plan:

- She plans her content every Sunday
- Checks sales and analytics every Friday
- Has an AI dashboard in Notion that updates weekly
- Uses ChatGPT for writing, planning, and monthly reflection

Result: She freed up 8 hours a week, doubled her client load, and said, "I finally feel like I'm running a business, not babysitting one."

Bonus Prompt: Hustla Mode Planner

Build me a customized planner for my business that breaks down daily, weekly, and monthly tasks using AI tools like ChatGPT, Canva, CapCut, Stripe, and Google Sheets.

Keep it simple, repeatable, and boss-ready.

AI Returns:

A plug-and-play template that becomes your **AI Business Operating System.**

The Hustla's Rule:

"You don't grow from grind alone; you grow from rhythm."

Let AI give you the structure to stay consistent, clear, and confident. Because once your habits match your ambition? **You're unstoppable.**

Peace Over Pressure: Mindset Shifts for Sustaining Success with AI

Let's get real.

You can have the best tools, the coldest content, and the tightest systems, but if your **mind ain't right**, it'll all feel like a burden instead of a blessing. This chapter ain't just about running your business. It's about **not letting your business run you.** Because we didn't start this journey to build digital prisons.

We built this to gain freedom. Time. Peace. Purpose.

And that means learning how to lead with intention, use AI with balance, and keep your **energy protected while your enterprise expands.**

Mindset Shift 1: You're the CEO, Not the System

AI is powerful, but it ain't the visionary. *You are.* You set the tone. You create the culture. You decide what matters. AI just helps you move faster. But don't let it rush you out of rhythm.

Use AI to reduce decisions, not to **replace direction**.
Use it to amplify your voice, not erase your soul.
Stay human in a digital world. That's your edge.

Prompt:

> *Give me a daily affirmation that reminds me I'm in control of my business, my time, and my energy.*

AI Might Say:
"I don't chase chaos. I command clarity. I build with intention and lead with peace."

Mindset Shift 2: From Hustle 24/7 → to Strategic Sprints

The grind is glamorized. But let's be honest, it's exhausting. The real flex is working smart, protecting your bandwidth, and designing your business to run even when you rest.

Automate 80% so you can be present for the 20% that matters: your creativity, your family, your health.

Prompt:

> *Help me design a 4-day workweek schedule using AI and automation. I want to protect my mornings, batch my content, and reserve Fridays for strategy or self-care.*

AI Might Deliver:

- Mon–Tues: Deep work + content
- Wed: Client calls + admin
- Thurs: Growth + visibility tasks
- Fri: No meetings. Journal. Reflect. Recharge.

That's not laziness. That's *longevity*.

Mindset Shift 3: From "I Gotta Do It All" → to "This System Got Me"

You were never meant to wear 12 hats every day.

If your to-do list still looks like it did before you found AI? You missed the point.

Delegate tasks to tech
Build SOPs (standard operating procedures) with AI
Train your tools like you'd train a team member

Prompt:

> *Create a simple SOP for how I post Reels using ChatGPT and CapCut. Include steps for writing, editing, and scheduling so I can hand this off to a VA in the future.*

AI builds it step-by-step so your business can move *without you micromanaging*.

Mindset Shift 4: Boundaries Protect the Brand

AI can help you grow fast. But growth without boundaries?

That's burnout.
Block time for yourself
Say no to things that don't align

Build auto-responses for emails, DMs, and inquiries so your peace don't get pulled every time someone asks for access

Prompt:

> *Write an out-of-office message that's professional, kind, and still lets people know I protect my time. I'll be offline on weekends but checking in Mondays.*

AI Might Say:

"Hey fam, thanks for tapping in. I'm currently off-grid to recharge and reset. I'll respond by Monday. In the meantime, check my FAQ or grab one of my digital products. Appreciate your patience and peace."

Now your time has a **tone** and people *respect it.*

Real-Life Example: Dre the Digital Strategist

Dre was scaling fast but burning out harder.

He used AI to:
- Build client dashboards
- Schedule content
- Write sales copy
  But he still felt overwhelmed.

Once he set clear AI-powered routines, work boundaries, and protected Fridays for self-care + planning?

His clarity increased

Client results improved

He fell back in love with the business

Dre realized: **Peace is part of the brand.**

Bonus Prompt: Hustla's Peace Protocol

> *Build me a daily mindset and business routine that protects my energy. Include a morning check-in, a mid-day pause, and an evening wind-down. Keep it spiritual, practical, and rooted in my role as a creative entrepreneur.*

AI Might Build:

- **Morning:** No phone first hour. Ask AI: "What's one thing I can do today to move with purpose?"
- **Mid-Day:** Stretch. Breathe. Ask: "What's draining me today? What can I delegate?"
- **Evening:** Reflect: "What did I build today? What do I need to release tonight?"

The Hustla's Rule:

"You are not your output. You are the architect."

AI helps you build, but it's on you to protect the builder. Let peace be part of the process. Because when your mind is calm, your business gets clearer, and your vision goes further.

Legacy Moves: What to Do Next and How to Stay Ahead

You didn't come this far just to stay in the same spot.
You flipped prompts into purpose.
You turned AI into a system.
You built clarity where there used to be chaos.

And you started running your business like a *visionary*, not just a hustla.

So what now?

This final section is about:

- Staying ahead of the game
- Cementing your systems
- Growing with purpose
- And building something **they can't cancel, copy, or compete with**

Because you've got the blueprint now. You ain't just working with AI. You're building **a legacy with it.**

Step 1: Evolve Your Systems
You've got your marketing, branding, and financial game automated. Now level it up.

Start documenting your AI systems like SOPs

Prepare to bring on help (VA, assistant, intern)

Use AI to train your future team

Turn your playbook into a course, workshop, or product

Prompt:

>*Help me turn my content creation process into a PDF guide I can give to a VA or sell as a digital product.*

Now you're not just operating, you're *educating.*

Step 2: Stay a Student (But Move Like a CEO)

AI is evolving *fast.* Every week there's a new feature, plug-in, or hack.

Set up Google Alerts for terms like "AI tools for entrepreneurs" or "AI in marketing"

Follow thought leaders in AI and business (or let ChatGPT recommend them)

Use ChatGPT to translate technical updates into "what this means for my business" language

Prompt:

>*Summarize the latest ChatGPT plugin updates and tell me how they could help me grow my coaching business. Keep it simple and practical.*

Now you're always ahead without being overwhelmed.

Step 3: Use AI to Build Generational Tools

You didn't learn all this just to *keep it to yourself.*

Use AI to help your kids write business plans

Create a family money tracker in Notion

Help your clients write their own systems

Leave *infrastructure* behind, not just income

Prompt:

> *Help me build a weekly entrepreneurship curriculum for teens in my community. Make it hands-on, rooted in culture, and infused with AI tools they can actually use.*

Now your knowledge feeds the block.

Step 4: Scale with Soul

You can grow without losing your roots.

Use AI to:
- Keep your brand voice locked as you scale
- Build systems before adding team members
- Forecast growth that doesn't cost your peace
- Run your business like a movement, not just a hustle

Prompt:

> *Create a 6-month growth plan for my apparel brand. I want to double sales, increase email subscribers, and launch a collab. Break it down by month with goals, content strategy, and financial checkpoints.*

Now your next chapter has structure *and* soul.

Real-Life Legacy Play: You, but Elevated

You already know what you've done:

You implemented, refined, and rebuilt.

Now here's what you do next:
- Revisit your AI tools quarterly
- Refine your prompts monthly
- Audit your business with AI weekly
- Recommit to your *vision* daily

Because success ain't a destination, it's a discipline.

Bonus Prompt: My Next 90 Days

> *Based on the systems I've built using AI, help me map my next 90 days of business growth. Include content focus, revenue targets, team-building steps, and self-care check-ins.*

AI Might Return:
- Month 1: Lock in funnel + relaunch lead magnet
- Month 2: Batch new content + build email list
- Month 3: Prep Q4 offer, start SOPs for VA, schedule rest days

Now you're building **consistently and consciously.**

The Hustla's Rule (The Final One):

"The future belongs to the ones who build with both intention and intelligence."

You ain't just surviving tech, you're *leading with it.*

Let the world keep guessing. Let them scroll and scroll. You?

You've got systems. You've got purpose. And now, you've got legacy in motion.

All eyez on AI. All power to you.

Let's build.

Conclusion

AI Ain't Just Tech It's Transformation

Let's stop and take this in...

Twelve chapters ago, you picked up this book looking for **tools**. Maybe you wanted to work faster. Post more. Make better content. Maybe you were just tryna figure out what all this "AI" talk was about. But what you walked into wasn't just a guide, it was a *gateway.*

Because *TrapGPT* was never about just technology.

It was about **taking the same mindset you brought to the hustle**, the work ethic, the resilience, the creativity, the community, and giving it a system that can *scale*.

It was about replacing burnout with **balance**, confusion with **clarity**, and "doing it all alone" with **doing it like a CEO.**

You didn't just learn prompts.

You learned **how to show up with power and peace.**

You Went from...

Winging your content →   To planning like a strategist

Guessing your numbers →   To analyzing like a CFO

Being everywhere →   To automating what don't need your touch

Talking about the brand →   To *being* the brand

You turned tech into *tools*. Tools into *systems*. And systems into **freedom.**

What This Journey Really Was

This wasn't about trends.

It wasn't about becoming the next AI guru.

It was about teaching you to:
- Take what you already know
- Combine it with what's evolving
- And use it to build something that lasts

Because truth is?

AI ain't the magic. *You* are.

AI just amplifies what's already inside you. And now that you've got this blueprint?

Nobody can take this mindset away from you.
Nobody can talk you out your worth.
Nobody can stop you from scaling unless you stop yourself.

The Hustla's Reminder:

"It was never just about the grind. It was about growth."

And now that you've tasted what strategy, systems, and soul-powered business can look like; **why would you ever go back to surviving?**

Next section coming with vision and heat. But for now?

Salute to you. You built different.

Don't Just Hustle... Scale.

So here you are. You got the blueprint.

You built the foundation.

You unlocked AI, not just as a tool, but as a **business partner, a strategist, a silent worker in your corner.**

Now it's time to do what real bosses do:

Step back, zoom out, and build something that can *outlive* you.

Because truth is... what you're building now; if you play it right, structure it tight, and scale it with soul; this could be the beginning of **real freedom.**

Not just a brand. Not just content. Not just a bag.

But a **legacy.**

Your Next Chapter Ain't About Doing More... It's About Building Smart

You already learned how to:
- Repurpose content
- Track your numbers
- Forecast revenue
- Protect your brand
- Manage your energy

Now here's the move:

Document every system

Train the tools until they sound like you

Build processes someone else could step into

Elevate from worker to *architect*

Prompt:

> *Help me turn my weekly business operations into a training manual or onboarding doc I can use for my first hire or future VA.*

Result: Your hustle just became **infrastructure**.

From Side Hustle to Enterprise Energy

You've been in creation mode. But now it's time to think bigger:
- What's your business look like with *five clients a day instead of one?*
- What if your systems could support *ten digital products?*
- What happens when your content brings in *daily sales without touching your phone?*

That's not fantasy.

That's **scale.**

And you're already halfway there.

Real-Life Blueprint: Tia the Educator

Tia started off selling $29 e-books. She used ChatGPT to write, Canva to design, and Gumroad to sell.

After reading this book, she:
- Built a 6-month content plan using AI
- Created a course with AI help
- Built a VA onboarding doc
- Launched a weekly AI-generated newsletter
- Used automation to run her entire funnel

Now she sells **freedom** not just PDFs.

> 5X'd her revenue
> Reclaimed her time
> Started licensing her content to schools and orgs

She didn't change her mission.

She **scaled her model.**

Bonus Prompt: Hustla's Expansion Plan

> *Based on what I've built so far, help me outline the next 12 months of growth. Include monthly themes, team expansion ideas, revenue goals, product opportunities, and rest periods. I want to scale without burnout.*

AI might deliver a full-year roadmap:
- Q1: Consolidate systems & launch 2 new offers
- Q2: Grow audience via podcast tour or Reels
- Q3: Hire VA & develop licensing opportunities
- Q4: Rest. Reflect. Relaunch.

The Hustla's Rule:

"Scaling ain't about doing more. It's about building smarter."

You already got the brand. The voice. The tools. The receipts. Now all you gotta do is **move like you've already arrived.** The vision is bigger than content. It's culture. It's community.

It's ownership.

Let's build it.

Trap the Tools. Free the Mind. Build the Legacy.

Look...

You didn't pick up this book to just learn some cool prompts. You picked it up because you knew you were capable of more.

More clarity.
More freedom.
More ownership.
More *impact*.

This AI game ain't about becoming a robot.

It's about becoming more *human* because now, the stuff that drained you? The stress, the guessing, the inconsistency? **It's automated.**

And in place of all that pressure?

You've now got **space to lead. Space to breathe. Space to** *build something real.*

The Trap Ain't Just Streets… It's the Cycle

This book is called *TrapGPT* for a reason.

Because too many of us got caught in the trap of:
- Grinding with no structure
- Posting with no plan
- Selling with no systems
- Creating with no energy left to *live*

But now? You different.

You turned the trap into a *tool.*
You flipped the cycle into a *system.*

And you learned how to **own your output without sacrificing your soul.**

You ain't just chasing opportunity anymore.
You're *building a lane so others can follow.*

So What's Next?

Next, you apply this daily.
Next, you teach somebody else.
Next, you use this to **own your time, your talent, and your territory.**

Maybe that looks like scaling your business.
Maybe it's mentoring young hustlas coming up.
Maybe it's turning this blueprint into a curriculum, a keynote, or a community.

Whatever it is, **don't keep this knowledge to yourself.**

Because the only thing more powerful than a boss who knows how to move?

Is one who shows others how to eat too.

Final Prompt: Hustla's Legacy Letter

> *Help me write a one-page letter to my future self or future kids, explaining why I built my brand with AI. I want it to reflect my struggle, my breakthrough, and what I hope they do with the tools I created.*

Because this ain't just for today, it's for the generation that comes after you. And when they ask how you did it? You can say, "I learned the system, mastered the tools, protected my peace... and then I passed it on."

The Final Hustla's Rule:

"The best thing you can automate is survival."

So you can finally focus on building what really matters.

AI ain't the shortcut. It's the shift.

You've seen the blueprint. You've run the plays. Now? **Lead. Own. Expand. Inspire.**

All Eyez on AI. All Power to You.

www.ingramcontent.com/pod-product-compliance
Lightning Source LLC
Chambersburg PA
CBHW080546220326
41599CB00032B/6375